COMMUNITIES, COUNCILS & A LOW-CARBON FUTURE

COMMUNITIES, COUNCILS & A LOW-CARBON FUTURE

What we can do if governments won't

Alexis Rowell

First published in 2010 by

Transition Books
an imprint of Green Books
www.transitionbooks.net

Green Books
Foxhole, Dartington
Totnes, Devon TQ9 6EB
www.greenbooks.co.uk

Design by Stephen Prior and Alix Wood. Cover illustration by Russell Hancock
Resources section compiled by Helen La Trobe
All photographs not otherwise credited are © Alexis Rowell

Back cover images:
Top: A vision of an urban housing estate. Photograph: Landlife
Middle: The school run by bicycle in Germany. Photograph: Alexis Rowell
Bottom: Open Space meeting at the 2010 Transition Network conference. Photograph: Mike Grenville
Page 2 image: The BTCV Carbon Army planting fruit trees on a London housing estate. Photograph: Chris Speirs

Printed in the UK by TJ International, Padstow, Cornwall
The text paper is made from 100% recycled post-consumer waste, the covers from 75% recycled material

ISBN 978 1 900322 65 2

CONTENTS

DEDICATIONS

This book is dedicated first and foremost to my partner Laura Watkins, who has had to put up with the vagaries of living life with an eco-warrior. She still methodically checks every leaf of my home-grown salad for bugs, she won't go anywhere near my 'yucky' wormeries and she's utterly terrified that I'm going to put a compost loo on the balcony. It's been a long hard ride for her, one where the rules change from day to day without warning, and it's far from over, but thus far Laura has stuck with me through thick and thin, and I love her all the more so for that.

It's also dedicated to my 'children' – Kieran, Dominic, Annabel, Adam, Megan, Roxy, Darcy, Lisiko, Noemie, Simon and Martin – who give me the strength and the rationale to continue fighting for their right to enjoy life on this planet as my generation has done.

Finally, I would like to dedicate this book to all the councils and community groups who have helped make it what it is and the future a little less bleak. So many of you are doing great things out there – we just need to help the others to do the same.

ACKNOWLEDGEMENTS

Thanks are due to Rob Hopkins, Sophy Banks and Naresh Giangrande for inspiring me to write the book; to John Elford and Amanda Cuthbert at Green Books for agreeing to publish (most of) my thoughts; to Alethea Doran for close-editing the text; to Sarah Nicholl for working tirelessly and selflessly with me to create and sustain Transition Belsize; to Cllrs Paul Braithwaite, Maya de Souza, Adrian Oliver, Penny Abraham, David Abrahams and Russell Eagling for being the backbone of the Camden Council all-party Sustainability Task Force; to Cllr Keith Moffitt, Leader of London Borough of Camden from 2006 to 2010, for letting me set up the Task Force; to Gloria Esposito, Ann Baker, Daniel White, Celeste Giusti and Dave Wilson of Camden Council for their hard work on this agenda and their willingness to keep me in the loop; to Philip Colligan, formerly of LB Camden, for his intelligence, humour and political savvy; to Rachel Stopard, Director of Environment at LB Camden, for her cheery, not-much-nonsense style; to Ed Watson, Head of Planning at LB Camden, and Harold Garner, of the Camden Sustainability Team, for eventually, hopefully, seeing the light on Passivhaus; to Oliver Myers and the LB Camden Sustainability Team for their hard work in making reality out of the Sustainability Task Force recommendations; to Quentin Given, Nusrat Yousuf, Kirsteen Harrison and Katherine Adams for the sterling efforts they made before LB Camden really started working on this agenda (which was after they'd all left); to Chit Chong, Liz Halsted, Martin Reading, Felicity Robinson, Lew Price, Richard Harris, Callum Johnson, Chris Nicola and Sam Monck at LB Camden for being pillars of good practice; to Jon Holland who contacted me shortly after joining LB Camden to see if we could chat about eco issues, thereby being the most proactive officer I met in all my time as a councillor; to Martyn Schofield, Simone Scott-Sawyer and Vinothan Sangarapillai for clerking the Sustainability Task Force; to Martin Black and Katherine Rees for keeping the Task Force website up to date; to Cllr Angela Mason for taking up the baton in May 2010 as the first-ever Cabinet Member for Sustainability at Camden Council; to Cllrs Paul Buchanan, Jacqi Hodgson, Linda Hull, Fi Macmillan and Robert Vint for sharing their stories; to Kate Dunlop and Sara Fakhro for their research work on the book; to Ben Brangwyn for his technological help with surveys and sensible suggestions; to Peter Lipman, Rob Hopkins, Ciaran Mundy, Jules Peck and Shaun Chamberlin for their work with me on Resilience Hustings; to my fellow Transition travellers in London, Duncan Law in Brixton, Lucy Neal in Tooting and Wendy Keenan in Haringey, and to Klaudia van Gool in Cornwall, for their love and support; to Suzy Edwards for her environmental work with me in Camden (and to Andy Williamson for keeping Suzy sane); to Mac, Kanada and Jonathan at Embercombe for opening my eyes; to Andrew Thornton for letting me help to green his Belsize Park Budgens and to his manager Jim McGuire for his patience and assistance; to Azul for providing Andrew Thornton with additional challenge; to Justin Bere and his architectural practice for their phenomenal help with Passivhaus; to Stephen Coleman for building a Passivhaus development in Highbury; to Sandy Halliday and Howard Liddell for being ahead of the curve on sustainable construction, for eliminating toxic materials from buildings and for introducing me to the concept of eco-bling; to Arthur Potts Dawson and Oliver Rowe for setting up eco-restaurants within walking distance of Camden Town Hall; to Nikki Packham, Iris Oren, Jess Gold, Jim Roland, Susan Poupard and Natasha Clayton at Camden Friends of the Earth, and Viv Stein at Brent FoE, for their campaigning zeal; to David Fleming, Lauren Thompson, Patrick Farkas, Debbie Bourne, Paul Mackay and everyone in Transition Belsize and the other Transition groups in Camden and North London for sharing the journey; to John-Paul Flintoff for proving that a man can and should make his own underpants; to Sam Hopley and Gavin Atkins at the Holy Cross Time Bank for their great ideas on co-production; to Farokh Khooroshi of the Fitzjohns Residents Association for pushing equally hard on everything and for his honey; to Jacob Tompkins and Joanne Zygmunt at Waterwise for their good advice; to water architect Jessica Read for being such an inspiration on waterscapes; to Lutz Johnen of Aquality for his help on grey water recycling; to Dusty Gedge of Living Roofs, Richard Sabin and Mark Laurence of Biotecture, and Matthew Frith for their visions on green roofs, green walls and greening housing estates; to Jane Riddiford for her Global Generators and their green roofs; to Matthew Thomson at the London Community Recycling Network for plugging away at third sector involvement in the solution; to Prof. Tim Lang, Dr David Barling and Dr Martin Caraher at City University for everything they taught me about food policy; to Kath Dalmeny, Ben Reynolds and all the folks at Sustain for their amazing work on sustainable food; to Patrick and Cathy Whitefield for their permaculture classes; to Angela Trisoglio for her help with 'meat and two veg', to Yvonne Rydin and Sara Bell at UCL for offering up their MSc students to do eco projects with Camden Council; to Ray Morgan at Woking, Adrian Hewitt at Merton, Thurstan Crockett at Brighton, Maxine Holdsworth at Islington, and Phil Webber and Helena Tinker at Kirklees for leading on this agenda and for being willing listeners; to Cllr Greg Foxsmith at Islington, Cllr Katrina Bull at Nottingham, and Cllrs Keith House and Louise Bloom at Eastleigh for being constantly open to new ideas; to Franny Armstrong for coming up with the 10:10 Campaign and to everyone on the 10:10 team for their hard work and good humour; to John Doggart for all his work on energy efficient retrofits; to Sarah Harrison for her eco-house and support; to Mathew Frith, Jacob Tompkins, Sheila Rowell, Shelagh Young, David Abrahams, Keith Moffitt, Sarah Nicholl, Rob Hopkins, Geoffrey Johns, Laura Watkins and Peter Lipman for reading through all or parts of the draft manuscript and making valuable comments; to Rob Hopkins (again) and especially Mike Grenville for helping with photos; to Cllr Tom Simon for filling in behind me in Belsize when I went AWOL to do this book; to our landlords Joe Elliot and Melissa Remus, and our ground-floor neighbours John and Laura Hardy, for taking on board all my eco-experiments uncomplainingly; to Dominic de Neuville for printing out 240 pages of final proofs as I passed through Zurich in August 2010; and to my mother for transcribing interviews and for still loving me despite my increasing eco-militancy.

FOREWORD

In late July 2008, the Transition Network office received a phone call from Somerset to tell us that the previous night Somerset County Council (SCC) had passed a remarkable resolution pledging the county's support to its local Transition Initiatives. It acknowledged the work of Transition Initiatives in Somerset, subscribed the Council to supporting the ethos of Transition, committed the Council to offering 'support and assistance' to those Initiatives, and committed SCC to becoming the 'first Transition Local Authority in the UK'. The caller asked, in the light of the resolution, whether we could tell them what a 'Transition Local Authority' actually is. We said we had no idea, but that we would be fascinated to explore it with them.

As Alexis points out so ably in this book, there are countless examples around Britain of councils taking bold and practical steps in response to climate change. Councils such as Woking and Kirklees have gone way beyond what is demanded by central government and shown bold leadership, making carbon reduction central to their policymaking. Some are taking inspirational initiatives in terms of energy efficiency in buildings, others are promoting cycling, and some are starting to give urban food production the importance it deserves. Yet there is not, so far, anything that might actually be confidently termed 'A Transition Local Authority' – no one authority that has begun to base all of its activities on climate change and peak oil, and on the need to strengthen local economies – although a resolution such as Somerset's is clearly a promising start.

From an early stage in the evolution of the Transition approach, we made it clear to anyone who asked that Transition is a process that 'local councils support but don't drive'. Our intention was to stress the point that Transition is intended to be a grassroots-led movement that communities themselves feel in charge of, and this has been one of the keys to the viral spread of the Transition concept. More recently, however, the discussion has moved on to what would happen if a community embracing the Transition approach, and vibrantly initiating a wide range of projects, engaged meaningfully with a local authority that has likewise begun to base its thinking and policymaking on responding to peak oil and climate change, and is seeking to make the area more resilient. What would it look like? That is one of the key questions addressed by this book. And what matters, as this book demonstrates, is that there are many people out there actually working to answer this question.

Much is said these days, especially as part of the new government's 'Big Society' agenda, about the concept of 'localism'. Localism, David Cameron's big idea, is the concept that political power needs to be returned to the local level, that government needs to be smaller and less omnipresent, and that any power that can be devolved to the community level is actually devolved.

Transition goes beyond this, to focus on *localisation*. Localisation is the idea that we try to shorten the distances between production and consumption as much as possible; that we strengthen local economies by striving to meet their needs from their local area. This is a key aspect of making communities more resilient to volatile oil prices and potential interruptions to supply, as important as actually reducing our carbon emissions – and in the current economic climate, I would argue, makes it more likely that communities will be able to weather future economic storms. Michael Shuman[1] describes localisation thus:

> ". . . it means nurturing locally owned businesses which use local resources sustainably, employ local workers at decent wages and serve primarily local consumers. It means becoming more self-sufficient, and less dependent on imports. Control moves from the boardrooms of distant corporations and back to the community where it belongs."

Government commitment to localism evokes cynicism in some. Lancaster City Councillor John Whitelegg told me in an interview that he was suspicious of politicians who used the term 'localism': "Britain is grossly over-centralised and I think that whenever a national politician starts talking about 'localism' their nose starts going into Pinocchio mode." However, the current focus on all things local offers a dynamic context for discussing many of the concepts and proposed projects that emerge from Transition Initiatives.

Alongside the work that some of our more innovative local authorities are doing, there is also some great work emerging from Transition Initiatives, as well as from the relationships that they are forming with their local councils. In May 2010 I went to the 'Unleashing' of Transition Malvern Hills. It was an amazing event. The group had taken a huge gamble and had booked the largest venue in the town, the local theatre, which seats over 400 people. With a week to go they had sold 100 tickets, and they were starting to sweat. In the end, the event was a sell-out, and a remarkable evening, with three choirs, a range of speakers and a palpable buzz of excitement and of possibility.

The highlight for me, though, was a section called 'Transition Endorsements'. Transition Malvern Hills' Will Tooby invited on to the stage 11 people representing key local organisations, who were asked to give a very short endorsement of Transition, or a reflection on the event. It was a dazzling assembly of the town's movers and shakers. The speakers included the leader of the local council, the principal of the local college, the Assistant Director of the local NHS Trust, the local police inspector, the head of the local Area of Outstanding Natural Beauty (AONB) and the newly elected MP for the area, Harriet Baldwin.

The County Councillor spoke of how Transition Malvern Hills (TMH) had supported their work in changing their catering arrangement: their food is now "not only edible, but it is all local!" Dave Armitage of the AONB, which has been supporting TMH's work, spoke of TMH's "great foresight and great energy", and urged the audience to "give them some of your energy and some of your time". The depth of support shown from such a diversity of key local organisations bodes very well for TMH's future, and Harriet Baldwin offered her support to the Initiative in whichever way she could help. This was a very strong foundation on which to build a well-linked and effective Transition Initiative.

Some councils are now starting to see the work of Transition Initiatives as representing a level of engagement with a potential that needs to be harnessed rather than brushed aside. Some Initiatives are working with their councils on very specific projects, and some are being supported with meeting-room space, grants, the support of council staff and so on.

Alexis Rowell came to Transition quite early in its emergence, and quickly saw its potential. He noted that Transition Initiatives needed to form meaningful relationships with local authorities and, in addition, needed people involved in Transition to put themselves forward for election. Alexis himself is a great example of the potential of doing so. What he managed to push through and to change within the Camden Council during his time as a councillor is quite extraordinary. The experiences he draws on from that time, as described in this book, are rich from the blood, sweat and tears he poured into this work. I can think of no one better than Alexis to draw together the many strands of the debates and discussions about how local authorities and Transition Initiatives can and do work together.

Interest in Transition from local councils continues to grow as the dynamism and visible impacts of Transition Initiatives similarly expands. I think that in many ways the work of Transition Initiatives is breaking new ground, asking questions that local authorities still struggle to address, and working with a freedom to experiment and to take risks that is rare in our current highly risk-averse culture. As the urgency to deal with climate change and peak oil becomes more widely understood, and as the realisation grows that a more resilient future lies not in the far-flung sourcing of globalisation but in the re-skilling and revaluing of local economies, it can't be too long before genuinely Transition local authorities start to emerge. The inspiration for their work, and the tools they utilise, will owe a great deal to this book, which makes a very welcome new addition to the Transition literature.

Rob Hopkins, September 2010

INTRODUCTION

It's very easy to fall into a vicious circle of helplessness, denial and despair over climate change and the inevitable end of cheap fossil fuels. Politicians struggle to take dramatic action when there's no visible crisis and a lack of any significant voter demand because they work on four- or five-year electoral cycles. At the same time governments do all sorts of things – like expand airports and build roads – that make it harder to meet distant carbon reduction targets.

Most businesses will take serious action, beyond saving money, only if they're forced to do so by legislation or if customers demand it. Meanwhile individuals and families are confused by mixed messages from the media and the government, or don't really understand why change is so urgent, or are unclear as to what they can do given the scale of the crisis, or simply find it easier to deny it.

I believe that councils, working hand in hand with local communities, can break the logjam and end the vicious circle. Bluntly, if the government will not or cannot or does not lead the way on the transition to a low-carbon future, then communities and councils can!

The idea that councils can play a significant role in saving the human race from itself is neither sexy nor fashionable, but the fact is, because they're obliged to think about the moral as well as the operational and business aspects of their actions, they could be a big part of the solution if they get better at opening up to new ideas and enthusiasm from the community.

Getting the message

I had my eco-epiphany in 2005. I read somewhere that the North Pole was going to melt in my lifetime. The news so shocked me that I went into a three-day depression. When I came out from under my duvet I decided I had to change my life. As a former BBC correspondent and international businessman I had a high carbon debt to repay.

My partner Laura and I learnt to recycle about 95 per cent of our household waste. Our food waste (and that from the flats above and below us) went into two wormeries and a compost heap that I surreptitiously positioned in our neighbour's garden. We ordered organic fruit and vegetables loose in a box delivered once a week. We refused plastic bags in shops. We more or less stopped eating meat and when we did we were really careful about where it came from in carbon and animal-welfare terms. We ate only sustainably sourced, small fish, usually from UK waters. I had water butts installed on our first-floor back balcony and in the neighbours' front yard to collect rain to water the fruit and vegetables growing on the balconies. I persuaded the landlady to let me grow more food in

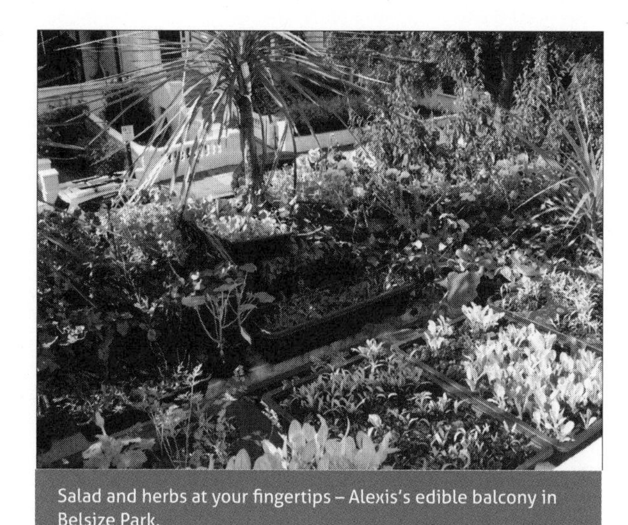

Salad and herbs at your fingertips – Alexis's edible balcony in Belsize Park.

the front garden in place of her ornamental shrubs. We cooked almost everything from fresh food and never bought processed meals.

We sold the car and bought bicycles. I stopped flying and Laura cut down on flights dramatically. We changed our electricity supplier to a renewable energy provider and our gas supplier to a biogas provider. In winter we put plastic sheeting on the windows and covered the gaps around our sash windows with masking tape. I put draught excluders on the doors and a balloon up the chimney. We gave up watching TV, thereby missing out on all the advertising designed to make you always want more, and we stopped shopping unless we absolutely needed something. I learnt to darn my socks and repair my underwear rather than throw them out.

None of these things was difficult and I would maintain that overall they have made the quality of our lives far better, but still the process was not easy, because I was always pushing for more; always testing the boundaries of what was feasible and acceptable.

The more I read about the science of climate change the more concerned I became about the future we were creating for our children and their children. I felt I had to do more. I had to try to raise awareness about the situation in the wider community. That led me to seek election in May 2006 as a local councillor in the London Borough of Camden.

I understood the significance of peak oil only after several years of eco-activism. I had pooh-poohed it, as the British government still does. But I eventually saw that I was looking through the wrong end of the telescope. All my life, oil has been running out. When I was at school the end of oil was 50 years off. It's probably still 50 years off or maybe 500 years off. The exact date is irrelevant. What matters is what happens after we go past the peak of supply. If demand continues to rise – and it's hard to see how it could do otherwise, given our never-ending demand for stuff and the development of the Chinese and Indian economies – then the price will rise and keep rising. But that price rise is likely to be increasingly volatile, including wild swings such as we saw in the summer of 2008. As many have pointed out before me, everything we depend on is based on cheap oil – our food, our clothes, our holidays, our consumerism; everything. What matters therefore is the end of cheap oil, not the end of oil.

Now, surely, I had a double-barrelled argument for action. There was the moral imperative to act on climate change for the sake of our children and grandchildren, and an economic imperative to act on peak oil to prepare for life after fossil fuels. I redoubled my efforts.

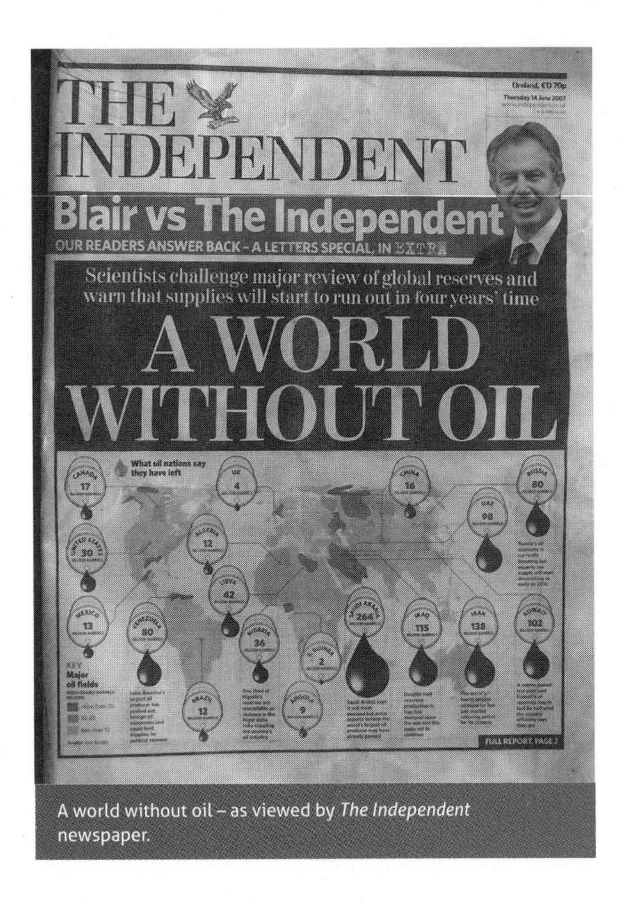

(Ireland, €1) 70p
Thursday 14 June 2007

THE INDEPENDENT

Blair vs The Independent

OUR READERS ANSWER BACK – A LETTERS SPECIAL, IN EXTRA

Scientists challenge major review of global reserves and warn that supplies will start to run out in four years' time

A WORLD WITHOUT OIL

What oil nations say they have left

Canada 17
UK 4
China 16
Russia 80
UAE 98
Algeria 12
United States 30
Libya 42
Mexico 13
Venezuela 80
Nigeria 36
E. Guinea 2
Saudi Arabia 264
Iraq 115
Iran 138
Kuwait 102
Brazil 12
Angola 9

KEY
Major oil fields

FULL REPORT, PAGE 2

A world without oil – as viewed by *The Independent* newspaper.

"Taken together, climate change and peak oil make a nearly airtight argument. We should reduce dependency on fossil fuels for the sake of future generations. But even if we choose not to do so because of the costs involved, the most important of those fossil fuels will soon become more scarce and expensive anyway, so complacency is simply not an option."

Richard Heinberg, author of ***The Party's Over***, ***Powerdown*** and ***Peak Everything***[1]

But two headless horsemen racing towards a fiery hell is no easier to sell than one, even for someone with a background in sales. I found people quickly tired of the message and the preacher, or preferred to ignore it. And my obsessive 24/7 approach meant I was at risk of burning out long before the flames would start to lick at anyone's feet.

There's an increasing body of research which says that the language of the apocalypse works only when the threat is immediate and personal. So, for example, the iceberg adverts of the 1980s were designed to instil so much fear in us about HIV-AIDS that we would all wear condoms. That worked. So did hard-hitting advertising campaigns about the increased risk of death or serious injury if you don't wear a seatbelt. But framing the problem in terms of disaster isn't working for climate change, and most people don't yet fully comprehend peak oil.

It seems that most people understand that there's a climate problem, but they're not being motivated to take action because the threat is perceived to be too far off or too difficult to solve at the individual level. Others confuse weather ('a cold winter means global warming isn't happening') and climate. Another factor in recent years has been the muddled messaging coming from the government (national carbon cuts enshrined in law versus airport expansion) and the media, which is always trying to create a polarised debate.

Undoubtedly we can all be clearer about explaining climate change and peak oil, in ways that make people psychologically more likely to take action. For some, slaying the dragon will work. That's clearly where climate change and peak oil campaigners are – swords in hands, smiting away 24/7. For others the message will need to be about cutting carbon pollution, ending energy waste, saving money and taking control of our lives back from energy companies who have us over an oil barrel.

For others it's perhaps about fun and being part of the in-crowd, although a 2009 World Wide Fund for Nature (WWF) study[2] rejected campaigning based on marketing, business cases and fashion values, which have helped create the problems. It suggested that it was impossible to effect lasting and meaningful change by trying to use existing values; arguing instead that only by changing underlying values and human identity can we succeed. In other words, there's too much 'surface framing' going on, e.g. 'ten easy steps to going green', or 'small steps, big difference', and not enough getting to grips with the big issues.

The conclusion the WWF researchers reached is that environmentalists need to:

- be aware of the coping mechanisms being deployed by people to avoid action, so that they don't trigger these mechanisms
- leave time and space for emotions to be expressed rather than jumping in with quick-fix solutions, which could themselves become a displacement coping mechanism
- be careful about not blaming certain groups and creating 'out-groups', e.g. 4x4 drivers.

The Transition perspective

It was when I discovered the Transition movement that I changed my tune. Suddenly I had something to say that went past the disaster-scenario planning of climate change and peak oil: I had a vision of a positive future to sell, one where the dragon has been slain and the princess has been won. Transition says "hang on a minute – sure we've got to do something about climate change and prepare for the inevitable end of cheap oil, but do we really want the same sort of society just without the carbon? Couldn't we imagine a better future, a greener future, a less-polluted future, less materialistic, less stressed, less chemical, healthier, happier; one where there is more time to enjoy the simple pleasures in life like good food and neighbours you can trust because you know them?"

Transition turns the problem of carbon on its head and asks communities to come up with a vision of a more positive future and then set a path towards that happier place. The solutions proposed by the Transition movement are the same as those needed to address climate change and peak oil – live within the natural limits of the Earth and without cheap fossil fuels – but the approach is far more inclusive and positive because it's about enhancing community.

Someone who's not interested in peak oil theory can still enjoy growing food. Someone who's unable to focus on climate change because its worst consequences will happen after they die can still appreciate an experience that brings together the community, such as a town meal or a harvest festival. Refocusing on the local, on your neighbours, builds social capital and creates more resilient communities. But what Transition also does is to create a space where people can be supported as they get to grips with the trauma of dramatic change. This psychological component both helps to keep people sane and provides a way in to the crucial question of how to achieve wider behavioural change.

One day in January 2010 I reflected on what I'd learnt from my local Transition group, Transition Belsize, over the previous week. I'd been (literally) blown away by a Draught Busting Workshop that took place on a Sunday afternoon in my flat. We learnt lots of low-cost techniques to make homes more energy-efficient and had a wonderful time working together as a community. On the Monday evening I joined the 'Sewing, Knitting, Mending and Making Gathering' of the Transition Belsize Arts and Creativity Group and learnt to mend my socks and turn up my trousers. On the Saturday I took part in a Bookbinding Workshop and was taught how to bind my own books using old cornflake packets. This is surely Transition at its best – an amazing combination of practical action and community spirit.

In many ways Transition simply repackages things that environmentalists, permaculturists and communitarians have known for years. But for me it was a new set of clothes. I was gripped because it put an incredibly positive spin on how to solve the problems of our age and create a better society in the process.

A holy task – Alexis learns to darn his socks at the Transition Belsize Sewing, Knitting, Mending and Making Gathering. Photograph: Sarah Nicholl

The role of local government

The Transition Handbook[3] by Rob Hopkins is very clear that local government is a critical element in the transition to a low-carbon future because of the levers it controls, such as planning, and that Transition groups therefore need to build bridges to local government. "You will not progress very far unless you have cultivated a positive and productive relationship with your local authority," writes Rob.

A key problem for community groups seeking to effect change in collaboration with a council is knowing how to work their way round the labyrinth that is local government. The main problem for councils is that the last three decades have seen power removed to Whitehall, and targets and performance indicators pouring in the other direction. Meanwhile, budgets have been under constant strain. It's rare to find many councils prepared to take risks or deviate from statutory obligations. However, it bears repeating many times over, and is one of the central themes of this book: it's truly amazing how many councils, often quite small district councils, are doing excellent work on this agenda above and beyond the calls of statute.

> "Local authorities in Britain still have significant freedoms – if they choose to use them. After many years of centralisation and the growth of a perception that councils have lost much of their freedom to act, it would be all too easy to imagine that local authorities had no capacity to make the kind of changes required as a result of climate change and/or the need to take action to reduce carbon consumption. In reality,

councils still have reasonable autonomy in the way they organise their own activities and also in their capacity to encourage particular forms of behaviour by their residents and businesses. Cutting carbon emissions is a sphere of activity where local government still has the capacity to lead and deliver."

Prof. Tony Travers, London School of Economics [4]

As Camden Eco Champion my remit was to scour this land and others for best practice on the environmental agenda in local government. Two things have struck me forcibly: there's a huge amount of good practice out there, emanating from both community groups and councils; and the leaders on this agenda are rarely doing what central government is telling them to do. Indeed, quite often central government is playing catch-up. That fills me with hope, because it clearly shows that councils and communities can make a difference if they want to.

But the majority of councils need help to understand why action is urgent, why the end of cheap oil is as important as climate change, how to frame the changes in terms of a different, more resilient, relocalised future, and how they can have much more influence than perhaps they realise. And that, hopefully, is where this book comes in – because I think councils and councillors have much to learn from community groups such as those that have sprung out of the Transition movement.

The best description of the rationale for this book came from Angela Ruffle of Sustainable Redland and Transition Bristol. "We need a 'how to' guide for a council chief executive," she told me, "as well as an influence manual for people who feel they're outside the fortress. And for people like me, who've plucked up courage to get inside, it needs to be a crash course in how to work within council culture." That's exactly what I've tried to do. Sometimes I'm talking to councillors and council officers; sometimes I'm talking to Transition Initiatives and other community groups. Hopefully, both groups will find things in this book that are useful.

I've taken climate change and peak oil as 'givens' in this book and concentrated instead on specific actions that community groups and councils can take. If you need more information about the likely consequences of peak oil for local authorities, then I recommend the 2009 Bristol Peak Oil Report[5] or the 2008 report 'Preparing for Peak Oil – Local Authorities and the Energy Crisis'[6] by the Oil Depletion Analysis Centre (ODAC), a UK charity, and the Post Carbon Institute, a US not-for-profit think tank. One of the authors of 'Preparing for Peak Oil', Daniel Lerch, has also written a book about climate change and peak oil aimed at local authorities in North America, called *Post Carbon Cities: Planning for energy and climate uncertainty.*[7] If you want to read more widely on the likely societal consequences of peak oil, then I recommend Richard Heinberg's books *Powerdown*, *The Party's Over* and *Peak Everything* or David Strahan's book *The Last Oil Shock* – see Resources section.

I've included examples of best eco-practice from local authorities and communities across the UK and elsewhere – to inform and, I hope, inspire councils and councillors, as well as local environmental activists, community groups and Transition Initiatives. Local government can be a huge driver for positive change, but rarely on its own. Communities need to understand what they can reasonably (and unreasonably!) ask for from local councils, and they need to know exactly what levers they can pull.

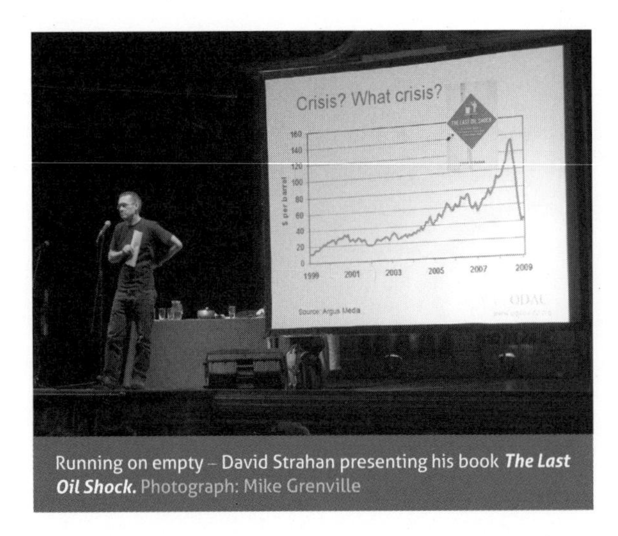

Running on empty – David Strahan presenting his book ***The Last Oil Shock.*** Photograph: Mike Grenville

Many people are wary of politicians. According to one survey in 2009[8] they were bottom of the list of trusted professionals, along with estate agents. I understand that, and it's one of the reasons why I'm so careful to walk the walk, to show by my behaviour that it's possible to do whatever I'm urging others to do. But I personally feel that we'll move a lot faster if we have people who understand the coming crises on the inside as well as the outside. It's one thing to invest time and effort into helping the existing power structures understand why it's so urgent to act now, but if activists can find ways to work within the power structures, then how much faster we would go. This book, therefore, also explains the ins and outs of getting elected.

There are now a few councillors who went into local politics because of climate change, peak oil or Transition. Their stories, their hopes and dreams, their achievements and frustrations, their learnings and conclusions, are recounted here.

It's not easy trying to effect dramatic change in a council as a backbencher. I've sometimes compared it with shifting the Titanic from its fateful course. Thankfully, many councils are starting to change direction, often at the behest of community groups. But there are more icebergs ahead. Too few councils understand the implications of the imminent end of cheap oil. Too few politicians walk the walk as well as they talk the talk.

> "Cornwall Council has commissioned a low-carbon strategy, which will be uncompromising in its recommendations. However, at the same time, they are running an airport and trying to expand domestic and European flights; giving waste contracts to firms known to favour incineration, encouraging unsustainable development, etc."
>
> Lindsay Southcombe, Transition Truro, Cornwall

Councils are not yet open enough to new ideas such as those coming from the Transition movement. But there are hopeful signs. In many cases councils, even quite small councils, are doing what's right for their community and the planet, even when external signals are telling them to do something else.

Sometimes that means taking risks. Sometimes it means spending money that local government increasingly hasn't got. But what choice is there? If central government will not or cannot do everything that needs to be done on climate change and peak oil, then local communities and councils will have to get on with it. And I would argue that we – and I speak as both a former councillor and an active member of my community – are better placed to do it.

> "We are now faced with the fact, my friends, that tomorrow is today; we are confronted with the fierce urgency of now; in this unfolding conundrum of life and history, there is such a thing as being too late. Procrastination is still the thief of time. Life often leaves us standing here, bare, naked, and dejected with a lost opportunity. The tide in the affairs of men does not remain at flood – it ebbs. We may not cry out desperately for time to pause in her passage, but time is adamant to every plea and rushes on. Over the bleached bones and jumbled residues of numerous civilisations are written the pathetic words, 'too late'."
>
> Martin Luther King[9]

Martin Luther King was speaking of another age when he made his "tomorrow is today" speech, but his words are, if anything, even more apposite at this stage of human development. We cannot wait. We need to act now. All of us. But the good news for those of us in local government and environmental community groups is that there's an enormous amount we can do. I firmly believe that councils and communities, working together, can make a huge difference on the climate change and peak oil agenda, as well as put forward a plausible vision of a better future and a route map to get us there. So let's get on with it – because I don't think anyone else is going to do it for us.

Chapter 1

A VISION OF A TRANSITION COUNCIL

Imagine a world where climate change is a reality. Of course that world already exists, but as far as we can tell we've not yet unleashed runaway climate change – a situation where the Earth tips into a highly unstable state and starts creating non-man-made greenhouse gases such that we lose control of the process altogether and life on Earth becomes untenable. We hope that runaway climate change will be averted by a combination of political will and individual action, but even if we do everything possible to avert disaster, average temperatures are likely to rise by at least 2°C.[1] That means there'll be more energy in the global weather system, which means there'll be more frequent extreme weather events. Melting ice caps and glaciers will cause sea levels to rise. Reduced rainfall will threaten freshwater sources and food supply chains. Millions of climate change refugees will leave home in search of less hostile climes. So we need to assume that the consequences of climate change will be dramatic.

> "Planet Earth, creation, the world in which civilisation developed, the world with climate patterns that we know and stable shorelines, is in imminent peril . . . Continued exploitation of all fossil fuels on Earth threatens not only the other millions of species on the planet but also the survival of humanity itself – and the timetable is shorter than we thought."
>
> Dr James Hansen[2]

Now imagine that the predictions of peak oil experts come true: that the halfway point in the world's oil supply is reached. Supply starts to fall, but demand continues to rise, and consequently prices go through the roof, making many of the things we take for granted around us impossible. So, for example, it becomes difficult and expensive to fly sugar snap peas from Zambia, or ship plastic toys from China, or buy apples from Australia or tomatoes from Spain, because all of those require cheap oil in the form of pesticides, fertiliser, plastics, aviation fuel, pumped water, shipping fuel, etc. We would then be facing a radically changed world and one that most people in the industrialised nations can barely contemplate.

> "Industrial civilisation is based on the consumption of energy resources that are inherently limited in quantity, and that are about to become scarce. When they do, competition for what remains will trigger dramatic economic and geopolitical events; in the end it may be impossible for even a single nation to sustain industrialism as we have known it during the twentieth century."
>
> Richard Heinberg[3]

Since the climate is changing and peak oil, indeed peak fossil fuels generally, is coming (in the sense that we will clearly reach the halfway point soon if we have not already done so), it makes sense to prepare for the next phase of human existence rather than stick our heads in the sand and wait for disaster to strike – which is what it feels like we're doing currently.

But here's another thought: without seeking to minimise the dramatic changes that are coming and their psychological effect on mankind, what if life can actually be better – more local, more natural, more friendly, less materialistic, less chemical, more trusting, less stressed, more healthy, more community-oriented – after oil? What if there is a better future out there waiting for us to go out and grasp with both hands? That doesn't mean it'll be easy to get there or to come to terms with the things we're bound to lose, but when we accept the changes, then as long as we prepare for them, surely a meaningful future is possible. And if communities and councils start working more resolutely together now, then how much easier the process will be.

All together now – the Chief Executive of South Hams Council at the launch of the Totnes Energy Descent Action Plan. Photograph: Rob Hopkins

Our brave new post-oil world will still need political structures. We'll still need local government in some shape or form. We'll still need councils, although they might change their names and remits, and we might change the way our representatives are elected. But what we really need from councils is for their vision to change. We need them to buy into the idea that life can be better after cheap oil and that they and local communities, working together, can play a significant role in the battle against climate change and in preparation for the depletion of natural resources. In short, we need them to help with a positive and practical vision of a society in transition.

What would a Transition council look like?

Since Transition councils don't yet exist it's impossible to say exactly what they'll look like. They're bound to come in a variety of shapes and sizes, since different regions have different priorities. Norfolk is projected to disappear under the waves if sea levels rise significantly. London is always nine meals away from starvation because it produces none of its own food. There's clearly no one right vision: there's only human ingenuity, willpower and energy. What follows are my thoughts, supplemented by suggestions from around the world, which I offer up as a basic outline to be refined by communities themselves.

Agreement on priorities and core values

In a Transition council both the politicians and the council officers would have a very complete understanding of climate change and resource depletion; both the global and local implications.

However, it's not enough to be well informed – a Transition council would have to have central organisational goals that relate to combating climate change, living within the planet's natural limits, preparing for the end of cheap oil and generally putting the well-being of human beings before that of developers, motorists, and any of the other interest groups that have done so much to put human society in such a precarious position. Every report that goes through a Transition council should be scrutinised to make sure that actions proposed use the minimum number of non-renewable resources and energy, and create the least possible waste or pollution.

> "A Transition council would have Transition built into its core values. All staff would have a commitment to the transition away from fossil-fuel dependency built into their job descriptions. All policies and all training would have Transition principles embedded. There would be a set of top-level objectives for the council with specific measures attached, relating to the carbon footprint of the city, food self-sufficiency, local energy production, jobs for a low-fossil-fuel future and low-carbon transport."
>
> Angela Raffle, Sustainable Redland and Transition Bristol

> "A Transition council would have a climate change team; would provide rewards for those who reduce their consumption of energy, water and waste; would work towards public events being carbon-neutral; and would plan public spaces for sea-level rise and increases in temperature."
>
> Debra Mill, Top End Transition, Darwin, Australia

Planning for a world without cheap fossil fuels

A Transition council, working with the community, would come up with a post-oil plan for its area – as Portland in the United States and Bristol in the UK have done. Bristol's Peak Oil Report is a fantastic template for community groups and councils wondering where to begin, and can be downloaded from the Bristol Council website.[4]

> "The local energy descent plan would be a foundation document of council."
>
> Robyn Hodge, Transition South Barwon, Geelong, Australia

In its 2008 report 'Preparing for Peak Oil: Local Authorities and the Energy Crisis'[5] the Oil Depletion Analysis Centre (ODAC), a UK charity, and the Post Carbon Institute, a US not-for-profit think tank, suggested that, based on a review of peak oil initiatives across the United States, Canada and Britain, local authorities should consider the following actions:

- Conduct a detailed energy audit of all council activities and buildings.
- Develop an emergency energy supply plan.
- Introduce rigorous energy efficiency and conservation programmes.
- Encourage a major shift from private to public transport, cycling and walking.
- Expand existing programmes such as cycle lanes and road pricing.
- Reduce overall transport demand by using planning powers to shape the built environment.

Key recommendations from 'Descending the Oil Peak: Navigating the Transition from Oil and Natural Gas', the 2007 report of the US City of Portland's Peak Oil Task Force[6]

1. Reduce total oil and natural gas consumption by 50 per cent over the next 25 years.
2. Inform citizens about peak oil and foster community and community-based solutions.
3. Engage business, government and community leaders to initiate planning and policy change.
4. Support land-use patterns that reduce transportation needs, promote walkability and provide easy access to services and transport options.
5. Design infrastructure to promote transportation options and facilitate efficient movement of freight, and prevent infrastructure investments that would not be prudent given fuel shortages and higher prices.
6. Encourage energy-efficient and renewable transportation choices.
7. Expand energy-efficient building programmes and incentives for all new and existing structures.
8. Preserve farmland and expand local food production and processing.
9. Identify and promote sustainable business opportunities.
10. Redesign the safety net and protect vulnerable and marginalised populations.
11. Prepare emergency plans for sudden and severe shortages.

Descending the Oil Peak:
Navigating the Transition
from Oil and Natural Gas

Report of the City of Portland
Peak Oil Task Force

March 2007

- Promote the use of locally produced, non-fossil-fuel transport fuels such as biogas and renewable electricity in both council operations and public transport.
- Launch a major public energy-awareness campaign.
- Find ways to encourage local food production and processing, and facilitate the reduction of energy used in refrigeration and transportation of food.
- Set up a joint peak oil task force with other councils and partner closely with existing community-led initiatives.
- Coordinate policy on peak oil and climate change.
- Adopt the Oil Depletion Protocol (an international draft agreement intended to mitigate peak oil by gradually and collaboratively lowering global oil demand).

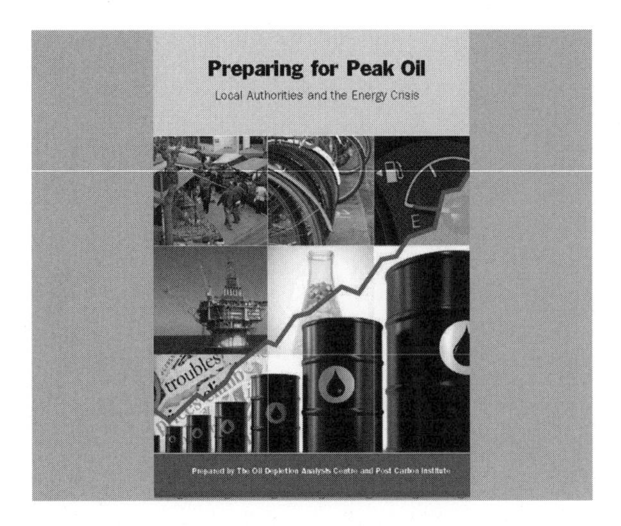

In 2008 the UK's Local Government Association (LGA) recognised its role in assisting Local Authorities with the consequences of peak oil and published its own report, 'Volatile Times: Transport, climate change and the price of oil'.[7] The LGA has also recognised the need for partnership with the Transition movement because of the need to spread awareness about relocalisation and local community resilience as well as developing local low-carbon economies.

The Improvement and Development Agency (IDeA), which supports improvement and innovation in local government in the UK, has a Low Carbon Futures arm that acknowledges the inevitability of peak oil. It highlights the work of the Transition movement and suggests that councils should work with Transition Initiatives and similar community groups.

> "No council can meet its targets without the help of the community. But councils should support from the sidelines and not try to take over. It's a slow process to build up the trust – it's got to come from the grass roots. Whether it's energy security, climate change mitigation or promoting social cohesion, there are 101 reasons why a local authority should support the [Transition] movement.
>
> It needn't cost a lot to help community groups. Sometimes just providing a venue for meetings or putting community groups in touch with each other is enough. Network Rail, for example, has been working with local food groups to make land available for growing organic food – disused railway embankments and cuttings are often ideal for this."
>
> Dr Jacky Lawrence, Strategic Energy Manager at Warwickshire County Council, interviewed by the IDeA[8]

More inclusive decision-making and openness to new ideas

Many residents complain that the councils are too bureaucratic and decisions are too remote. Of course council officers need rules, structure and targets if they are to work coherently and efficiently. But they also need to be open to new ideas. Transition councils would pay close attention to the work being done by community groups on inclusive decision-making systems such as Open Space or World Café.

Open Space is a powerful tool for engaging large groups of people in discussions to explore particular questions or issues.[9] It can be used with groups of anything between 10 and 1,000+ people. It shouldn't work but it does!

An Open Space event needs a big open question to kickstart it, like 'How will Glasgow feed itself beyond the age of cheap oil?' In the first part of the event participants are asked to come up with headline answers to the big question of the day, e.g. 'By growing more food.' Those then become the themes for breakout groups. At the end, the different breakout group leaders sum up their discussions for the whole group.

Open Space events will almost always be fun and lead to an amazing number of ideas being generated and noted down, as well as making everyone feel thoroughly part of the decision-making process.

World Café is like a bounded version of Open Space, in that the questions to be discussed are framed in advance by the organisers. So, for example, Transition Town Totnes held a World Café with candidates in the run-up to the May 2010 local and general elections. The candidates were ranged around the room at tables and told to expect questions from voters on resilience issues.

To my mind these are ideal techniques for councils seeking to engage with residents.

Whither the Transition movement? An Open Space meeting at the 2010 Transition Network Conference.
Photograph: Mike Grenville

A Transition council might use the Four Rules of Open Space and the Law of Two Feet

The Four Rules of Open Space:

1. Whoever comes are the right people.
2. Whatever happens is the only thing that could have happened.
3. Whenever it starts is the right time.
4. When it's over, it's over.

The Law of Two Feet:

'If, during the course of the gathering, any people find themselves in a situation where they are neither learning nor contributing, they must use their feet and go to some more productive place.'

"In its governance processes a Transition council would highlight and facilitate engagement with community and offer best practice in access to council information. It would frequently conduct open forums and actively invite the public into those, especially with regard to assisting the council in learning how to become more sustainable and useful itself, but also to resolve curly questions."

Chris Harries, Waterworks Valley,
Tasmania, Australia

"A Transition council would be participatory. The radical transition expected requires mass participation in decision-making at all levels. Think of it as the end of representative democracy, elites and experts. The idea of distinct Transition Groups and Transition council is nuts: Transition thinking is spreading faster and in more diverse and interesting ways than can be driven by Transition groups."

Bob Thorp, Transition Keighley, West Yorkshire

"I hope a Transition council would evoke massive interest locally, instead of the present cynical attitude most people have toward their councils. Our Town Council has re-elected itself for years without actually having elections and the Transition activity ought to correct this."

Michael Dunwell, Transition Forest of Dean, Gloucestershire

"A Transition council would work together with its local Transition group. It would understand our passion and need for community building. It would not be separate and apart. It would understand that we are in the community together."

Pupak Haghighi-Brinch, Transition Forest Row, East Sussex

In a Transition council decision-making would be decentralised down to the lowest possible level, possibly even to street level. Many councils have introduced Local Area Forums or Committees as a way of listening more carefully to communities, and some have been given resources to spend as they see fit.

"A Transition council would work with local Transition groups but not be part of them. If representatives of a Transition council asked for our input, provided some funding and attended our public meetings, then that would be a great start."

Klaudia van Gool, Transition Liskeard, Cornwall

"A Transition council would be open to community-led ventures and self-organising projects. The council would not employ its 'own' staff to run projects (who then spend half their lives in meetings) but use its funding to devolve delivery."

Linda Screen, Transition Town Dorchester, Dorset

"I think the biggest challenge is to keep the ideas coming in from the bottom up, rather than getting into the usual top-down model where we are told what is being done in our own best interests, instead of establishing what those are for ourselves. Ideally a Transition council would facilitate rather than lead."

Sophie Galleymore-Bird, Transition Rame, Cornwall

One should never forget that the council is made up of individuals with their own hopes and fears. At the moment there are artificial divisions between councillors and council officers, designed to protect the council from party politics. Officers can seem to be remote from the communities they serve and often don't live in the borough they work for. In a Transition council officers would be socially closer to the community and their representatives, and ideally officers would live in the community they work for.

"A Transition council would understand that it's not a machine but a collection of individuals who are also part of the community – real people with real feelings, fears, inspirations, hopes. A Transition council would value contributions and initiatives from all its staff right across the board on sustainability matters. It would be open to partnership working with the community, rather than having a them-and-us mentality (which also means having a community that sees the council as a partner, rather than harbouring their own them-and-us stuff). A Transition council would be politically brave in supporting any initiative that builds resilience in the face of nimbyism. It would use incentives to build resilience in an imaginative way. It would invest – and support the community in investing – in local renewables. It would celebrate what we're all doing.

Chrissie Godfrey, Transition Town Taunton, Somerset

Enabling and broadening community action

A Transition council would be an enabling force. Energy and ideas would come from residents and local communities, but assistance in terms of structure, purchasing, planning and other activities that cannot be done at an individual or a very local level would come from the council. So, with food growing, the example of Middlesbrough is instructive. There the council has set itself up as an adviser to local people, as a negotiator on behalf of residents, and as a teacher of food-growing skills to local people. If a resident identifies a piece of land that could be used for food growing, then the council will find out who owns the land, negotiate with the landowner on behalf of the residents or residents' group, and advise what can be grown and how. This is a Transition council in operation – even if Middlesbrough didn't really set out to be that.

"A Transition council brings local strategic partners together to lead by example on tackling climate change and peak oil through visible manifestations such as solar thermal panels on public buildings, green roofs, green vehicle fleets, community gardens of council lawns, interactive sustainable schools that act as educational tools. The council would work with and enable community groups, including Transition groups, to help develop bottom-up community ownership and responsibility for addressing climate change and peak oil. The council should foster a culture of collective responsibility rather than a 'them-and-us' culture that persists in many areas. There should be a comprehensive programme of support of Transition groups and others doing similar work: grants, training, revolving loan funds, etc. to help empower green initiatives."

Denny Gray, Transition Town Wandsworth and Wandsworth Environment Forum

"A Transition council should enable our ideas to become reality, things like: eco-homes for social tenants and new homeowners; support for renewable power and community heating systems; help for local food to be produced, exchanged and sold; support for green industries and training in green jobs; safe cycle and walking networks; affordable and reliable and regular public transport, with room for bikes to be carried; and support for car-share schemes."

Susan Westlake, Transition Tunbridge Wells, Kent

"A Transition council would provide no-cost facilities to Transition Initiatives, such as a place to meet, and include familiarisation with the Transition idea in its sustainability education programme; would include Transition input, as well as other public input, in deliberation for policy formulation; would develop some form of collaboration with Transition Initiatives on relevant topics; would demonstrate Transition practices in procurement, e.g. energy production/ purchasing."

Russ Grayson, Transition Sydney, Australia

Leading from the front

The chief executive of a Transition council would clearly need to understand the urgency of climate change and peak oil, and be prepared to put what is right for the local area ahead of what central government may wrongly require. Many chief executives have plied their trade through a long dark winter of centralisation and outsourcing of services masquerading as minimisation of waste (1979-97) and, more recently, endless performance indicators dressed up as increased productivity (1997-2010). There was precious little reward on offer for those who took risks. But that period is surely now over. The only way for power to go is down – the new coalition government in the UK included plans for a Decentralisation and Localism Bill in the Queen's Speech in May 2010. And the evidence of this book is clear – local government is starting to do things that are right for local communities and the planet rather than being dictated by central government.

Central government has been good at setting long-term emissions reduction targets but less good at taking immediate action such as curtailing airport and road expansion. That looked like changing with the new UK coalition government elected in May 2010, who announced that plans for the expansion of Heathrow, Gatwick and Stansted would be shelved, although it didn't rule out expansion elsewhere and there was no sign of the road-building budget being cut back.

Parliament and Whitehall may yet come through with the right approach and the right solutions, but the chief executive of a Transition council would understand that we cannot afford to wait for central government, just in case they continue to be unable to make the right hand and the left hand work together. We need to get cracking at the local level and take control of the solution for ourselves.

The political leader of a Transition council would be clear about where we are going and use every opportunity to show that, if we prepare for it together, if we are bold, if we accept that dramatic changes will have to be made by all of us, then the future can be a better place, or at least a perfectly acceptable place. This positive vision of the future is critical if we are to deal with the extreme psychological issues associated with dramatic change.

More elected eco-activists

A Transition council would have more elected members who can not only deal with the bread-and-butter issues – the cracked pavements and dog mess of their ward – but can also have one eye to the immense changes of the future, and understand what needs to be done both at a local and global level.

"A new generation of councillors is probably needed who have greater awareness of global issues and how they might relate to the local community. Somehow the political fixation on local economic development and short-term job creation needs to be channelled in the Transition direction because the current paradigm is dominated by Tesco-thinking and retail-led regeneration."

Sean Furey, Protect Kent (supporting environment groups around Kent & Medway)

> "A Transition council would be made up of people who have patience and are thick-skinned!"
>
> Andrew Capel, Energy Descent Llanidloes, Wales

Measuring human progress, impact and community differently

At the moment most councils have a legal and financial sign off for all reports, but a Transition council might make happiness, resilience and ecological footprint key hurdles that every decision has to cross. As a society we are no happier today than we were in the 1960s.[10] By some measures we are even less happy. A Transition council would stop focusing on meaningless economic statistics and the accumulation of stuff, get real about what makes us happy, and plan council expenditure and policies around that.

> "A Transition council would seek to ensure the resilience of the county. For example, by fully supporting community energy initiatives and a local currency backed by the energy produced. There would also be a balance between the value given to personal well-being and the economic worth of the population."
>
> John Marshall, Transition Helston & District, Cornwall

Caerphilly Council in Wales has been working with the new economics foundation on ways to measure happiness, health and ecological footprint. The London Borough of Sutton is trying to act according to One Planet Living principles, which were devised by WWF and BioRegional (see box overleaf). These are steps in the right direction.

A Transition council might also measure resilience by measuring the following.

- Percentage of food grown locally.
- Amount of locally issued currency (e.g. Brixton Pounds) in circulation as a percentage of total money in circulation.
- Number of businesses locally owned.
- Average commuting distance for workers in the town.
- Average commuting distance for people living in the town but working outside it.
- Percentage of energy produced locally.
- Quantity or proportion of building materials made from recycled resources.
- Proportion of a representative basic of essential goods being manufactured in the community within a given distance.
- Proportion of compostable 'waste' that is actually composted.
- Number of children who have held a frog or worms.

Adapted from the *Transition Initiatives Primer*[11]

A Transition council would be continually asking if there were a less resource-intensive way to carry out a particular policy: one that minimised production of waste or pollution, and one that human beings would be more likely to enjoy. At the moment we don't know

BioRegional/WWF Principles of One Planet Living as adopted by the London Borough of Sutton[12]

1. Zero carbon – our climate is changing because of human-induced build-up of CO_2 in the atmosphere.
2. Zero waste – waste from discarded products and packaging creates disposal problems and squanders valuable resources.
3. Sustainable transport – travel by car and aeroplane is contributing to climate change, air and noise pollution, and congestion.
4. Local and sustainable materials – destructive resources exploitation (e.g. in construction and manufacturing) increases environmental damage and reduces benefits to local community.
5. Local and sustainable food – industrial agriculture produces food of uncertain quality, harms local ecosystems, and may have high transport impacts.
6. Sustainable water – local supplies of fresh water are often insufficient to meet human needs, due to pollution, disruption of hydrological cycles and depletion.
7. Natural habitats and wildlife – loss of biodiversity due to development in natural areas and over-exploitation of natural resources.
8. Culture and heritage – local cultural heritage is being lost throughout the world due to globalisation, resulting in loss of local identity and knowledge.
9. Equity and fair trade – some in the industrialised world live in relative poverty, while many in the developing world cannot meet their basic needs from what they produce or sell.
10. Health and happiness – rising wealth and greater health and happiness increasingly diverge, raising questions about the true basis of well-being and contentment.

www.oneplanetliving.org

the answers because it's nobody's responsibility to ask those questions.

Space to deal with the psychology of behaviour change

The leaders of a Transition council would agree that it's possible to have a positive future despite climate change and peak oil if we prepare physically and psychologically for it. They would create the space for achieving change and dealing with reaction to change.

In late 2009 the number of people expressing doubt about human-induced climate change started to rise in the UK and the US. Scepticism among Americans split along party lines, with Republicans arguing that climate change was a conspiracy designed to end USA hegemony and overturn the American dream. In November 2009 George Monbiot wrote a compelling piece about how many UK sceptics were people close to the end of their lives.[13] It could also be that the sceptics are struggling with what many psychologists call the grief cycle (see box, opposite). This is about trying to understand how people react to grief and the

Opening a can of worms – pupils at Camden's Holy Trinity Primary School get to grips with their new wormery. Photograph: Polly Hancock

stages they have to go through before they can accept what has happened.

All of this can be used to interpret people's reaction to climate change and peak oil. It can also be used to think about how institutions and businesses might understand the impending demise of their business models in a world without cheap oil. It may not be a linear process. Building on Kübler-Ross, William Worden explained grief as a series of tasks that could be embraced or refused, tackled or abandoned, but that would never be complete. Meaning may be restored to one's life and joy may again be possible, but the work may also falter and stall, depression may temporarily return, and it may take some time before the work continues.[14] This may well be what happens with our reaction to changes as dramatic as climate change and peak oil.

The grief cycle

The Kübler-Ross model, commonly known as the five stages of grief, was first introduced by Elisabeth Kübler-Ross in her 1969 book *On Death and Dying.*[15] It describes, in five discrete stages, a process by which people deal with grief and tragedy, especially when diagnosed with a terminal illness or catastrophic loss.

Shock

The first reaction is usually one of shock and complete disbelief. Many people quickly experience complex and confused feelings – anger, guilt, despair, emptiness, helplessness and hopelessness.

Denial

When the shock begins to wear off, many people go through a stage of denial during which they cannot accept the reality of the loss.

Anger and guilt

It is common to experience anger, sometimes guilt and often both. Many people find themselves asking: 'Why has this happened?' 'Why me?' This is particularly so if the loss was sudden, unexpected or involved a tragic accident.

Despair and depression

In the first few weeks the whole situation may seem unbearable and in the months that follow many people feel there is little purpose in life and nothing of interest in the outside world.

Acceptance

Eventually people pass through the period of depression and begin to accept the loss. This usually happens with the passage of time and, as the pain eases, we are able to think about our loved one and recall the past without feelings of devastation.

According to Sophy Banks, one of the founders of the Transition movement, the grief cycle is just one component. She argues that to fully understand the psychological effects one needs to explore: why, as a society, we are addicted to what we have; why there's a gap between what we think we need and what we really need; how to deal with loss and traumatic change using variants on the grief cycle; and how identity can potentially shift – say from climate sceptic to hater of big-car drivers – very rapidly during the process.

Above all, she says, there must be a complementary process of visioning the future in positive terms:

"The Transition movement puts out a distinct message that it is possible to have an abundant, vibrant, inclusive and sustainable future way of life, and that to do so we need to vision and build it now. If you read some of the materials that are out there to encourage people to engage you will find this strongly emphasised. Why is this? Because to engage at all with the

Eyes wide shut – Sophy Banks leads a visionary moment at the 2010 Transition Network Conference. Photograph: Mike Grenville

> enormous and terrifying information many people need something positive to balance the fear. Otherwise apathy reigns, not because we don't care but because we are cut off from feeling our love for life on Earth when we feel it profoundly threatened and believe ourselves to be powerless to do anything. The scale of the problem is overwhelming and unmanageable, especially on my own.
>
> The message of Transition – that together we can face this and create an enjoyable, living, healthy world is both plausible and empowering – and has resulted in many thousands of people around the world engaging with the issue. And I believe that both pieces of this message – that we do it together, and that there are many good things to be had as a result – are crucial to making engagement possible."
>
> Sophy Banks, Transition Network[16]

A Transition council must engage with community groups on all these psychological issues if it is to understand what creates behavioural change and to help residents to come to terms with dramatic change. Our cultural stories and social norms will have to change if we are to cope with climate change and peak oil. A Transition council will have to create the space for psychological work if its work is to have lasting change.

In brief . . .

So, in summary, a Transition council would be one where politicians and officers are fully aware of the implications of climate change and peak oil; where the community and council have a plan for life after cheap oil; where the council acts as an enabling force for community groups and as a liberator of new ideas from the community; where the council's leaders put the local area and the planet before the needs of national government wherever necessary; where the priority is on increasing the happiness or well-being of residents above all else; and where there is space for a deeper understanding of the dramatic changes we are going through and how to achieve behaviour change. It would also agree that it's possible to have a positive future despite climate change and peak oil, if we prepare physically and psychologically for it.

> "What we have done, we can undo. There is no longer time to waste nor any need to accumulate more evidence of disasters; the time for action is here. I deeply believe that people are the only critical resource needed by people; we ourselves, if we organise our talents, are sufficient for each other. What is more, we will either survive together, or none of us will survive."
>
> Bill Mollison, co-founder of the permaculture concept[17]

Chapter 2

FINDING YOUR WAY AROUND LOCAL GOVERNMENT AND INFLUENCING A COUNCIL

Questions answered in this chapter:

- How is local government structured?
- What does the jargon mean?
- How can you get involved with emergency or resilience planning?
- What legislation might be useful to community groups?
- How can community groups influence councils?

How is local government structured?

Local government can be a confusing labyrinth. It took me a good year to find my way around and even after four years there were parts I knew next to nothing about. A recurring theme in this book is the need for patience when engaging with councils. As one wise old lag said to me early on: "It's a marathon, this business, not a sprint!" Who you need to talk to depends on what exactly you want. How far you get depends on what you bring to the table and what your approach is.

Principal authorities

Some parts of England have a two-tier structure – a county council responsible for regional services such as education, waste disposal and strategic planning, and several district councils responsible for local services such as housing, waste collection and local planning. Some areas – known as unitary authorities or metropolitan borough councils – have only one level of local government and do everything. London boroughs have a regional layer of government above them – the Mayor of London and the Greater London Authority – which is responsible for some strategic services such as transport. Wales, Scotland and Northern Ireland are made up of unitary authorities.

In 2007 the Scottish government signed a concordat with local authorities promising not to tinker with the structure of local government during the then Scottish Parliament, setting out guaranteed levels of funding throughout the parliament, giving councils more discretion over how to spend their funding, streamlining targets into one set of performance indicators and setting Single Outcome Agreements with each authority based on those performance indicators and national objectives. Scottish councils cooperate through, and are represented collectively by, the Convention of Scottish Local Authorities (COSLA).

The Local Government Act 2000 required councils in England and Wales to move from a committee-based system of decision-making to one with an Executive, with either the Council Leader and a Cabinet acting as an Executive authority, or a directly elected Mayor. The Act allows for local referendums to decide whether to have a Mayor. By mid-2010, 30 of these had been held; the majority deciding against the elected mayor option.[1]

The Executive, in whichever form, is supposedly held to account by the remainder of the councillors via scrutiny committees. In practice the Executive has virtually all the power, much like the British government does with respect to Parliament. Backbench councillors can change things, but usually only if they are prepared to put a lot of energy into organising cross-party campaigns involving residents and the press.

Parish and town councils

Below the district or unitary level, an area may be divided into parish or town or (in Scotland) community councils. Typical activities undertaken by a parish, town or community council include allotments, parks and public clocks. They also have a consultative role in planning. Parish, town and community councils have traditionally existed in rural areas, e.g. villages or small towns. However, in 2007 the law changed to allow parish councils to be formed in urban areas. They already exist in Birmingham, Leeds and Newcastle. In the London Borough of Barnet residents in East Finchley have been trying to create one (see box).

As a Londoner I had little knowledge about, or interest in, parish and town councils when I started writing

Forming a local council in London

In the East Finchley neighbourhood of the London Borough of Barnet a group of Transitioners became so fed up with the local council removing speed humps and cycle lanes that in the autumn of 2009 they started looking into the possibility of establishing a local council.

The Local Government and Public Involvement in Health Act 2007 devolves the power to take decisions on matters such as the creation of parish councils to principal authorities (a district, borough or unitary council). Under the previous legislation decisions on the creation of parish councils lay with the Secretary of State. The intention of the new legislation was to simplify the process and to make it more local.

Under the Act, if 10 per cent of a particular community is in agreement, then an application can be made for the establishment of a new local council with boundaries defined by the applicants. Its powers are equivalent to a parish or town council devolved down from the principal authority, which, in the case of East Finchley, would be Barnet Council. With help from the National Association of Local Councils the group began consulting the residents of East Finchley in the spring of 2010.

The need to persuade 10 per cent of the community to agree is quite a hurdle – in Potters Bar, just up the road, where residents also tried to set up a local council, they couldn't persuade enough of their neighbours that it was a good idea. Votes were sent to all 16,443 Potters Bar residents on the electoral register in October 2009 and 44 per cent were returned. 4,446 voted against, 2,807 voted in favour and the scheme was abandoned.

East Finchley's Transitioners knew when they started that it would take a long time to achieve their dream, if they ever did, but whatever happens they say one useful by-product has been the opportunity the process has provided to bring together different local groups to discuss issues of mutual interest such as transport, food and the environment.

this book, but I increasingly see that they can be more important than I had thought, especially in terms of allotment provision (although not in central London, where boroughs do not have to provide allotments) and planning consultation.

> "In a village community rather than in a town community you experience things differently. Whilst it's true that on the mechanical level parish and town councils don't have the august powers of the district or county council, they are the conduit to them and a strong local influence."
>
> Mike Grenville, Transition Forest Row, East Sussex

> "Parish and town councils have two key links with the principal regional authority – they have a statutory route into planning and therefore an incentive to acquire expertise about planning issues. Also, if the unitary authority sees a town or parish council working collaboratively with a Transition or community group, then it is more likely to see it as a suitable conduit for external funding. The advantage of working at the parish or town council level is that we can do whatever we want so long as we can raise the funding for it. We're not bound by thousands of statutory duties as boroughs are.
>
> Parish councils and Transition groups are probably the two organisations best placed to achieve something together. I rely on the work the Transition group is doing, to support the arguments, to make the case, to provide the information and the commitment, the physical manifestations of the project, because that suggests they are aware of what is possible in their own community and they look to town/parish councils to say 'come on folks, get off your arse, we can demonstrate that we've got support, we can demonstrate that this is an issue, we've got some of the working solutions, we understand how this might work going into the future alongside all the arguments on threat and exposure and what we face, why not work with us?'"
>
> Allan Bosley, Chair of Corsham Parish Council, Wiltshire

Funding

Local councils are funded by a combination of central government grants, council tax (a locally set charge based loosely on house values), business rates (which are set nationally, collected by councils, sent to central government, and then redistributed in grants to councils), and fees and charges from certain services including parking enforcement. Only 25 per cent of local government funding comes directly from council tax, which means that if a council wishes to increase its funding modestly, it has to put up council tax by a large amount. Central government retains the right to 'cap' council tax if it deems it to be too much. (In an extraordinary display of heavy-handed centralism by an administration apparently wedded to localism, the new coalition government in the UK announced soon after it came to power in May 2010 that it was freezing council tax for everyone in England and Wales for a year.)

Council tax is collected by the district-level council. Authorities such as the Greater London Authority, parish councils, county councils, passenger transport authorities, fire authorities, police authorities and national parks authorities can apply a precept, which means add a charge. This shows up as an independent element on council tax bills, but is collected by the district and funnelled to the precepting authority.

Councillors

A councillor is elected to represent the people of a particular ward (electoral district) for four years. There can be up to three councillors per ward. Wards in a council area are roughly the same size in terms of population, so that each councillor is responsible for a similar number of residents. Being a councillor is paid as a part-time job, and sometimes only enough to cover expenses, although many councillors spend their entire working week doing it – and often their weekends too!

There are as many reasons why people seek election into local government as there are councillors. Some people are genuinely motivated by representing their local community or changing things for the better. Others see it as a stepping-stone to higher politics. Some have a particular issue they want to push. Others see it as a way of raising their status in their own particular community. When you find out what makes an individual councillor tick, then you'll know how to approach that particular person when you need help or you want to move things forward.

> "When we first started, the parish council was very resistant – people would say to my face 'what right do you have to call us a Transition Village?' We've kind of co-existed with each other. We try to include them, we try to engage in supporting them, such as when they wanted to enhance the monthly community market, and we've presented to the parish council meetings what Transition means. I think the Forest Row Energy Descent Plan, which we've published and delivered to every household in the spring of this year, was very professionally done and I think that helped – our standing improved as a result of it. So I try to encourage people to continue to have dialogue with them, but not to get stuck in feeling that we need to have their approval to do anything, but to keep the doors open."
>
> Mike Grenville, Transition Forest Row, East Sussex

Council meetings

Full Council meetings are attended by all councillors. They take place about every six weeks or so and are less important than they look. They're set pieces that have little power other than to pass the annual budget or agree appointments to committees (not officer appointments) and who should be Leader of the council. That said, there are three elements within council meetings worth knowing about that can be helpful to a community group's cause: deputations, questions to Executive or Cabinet Members and motions. All councils have their own constitutions so they may call these elements by different names, or they may have variations on them, but most will have them in some shape or form.

Waste not want not – Sustaining Dunbar energy-audits its local councillors. Photograph: Philip Revell

Deputations

A deputation is where a person or group of people asks to raise an issue they're concerned about at a meeting of the council. To make a deputation you need to contact the relevant committee clerk in writing. You can find the clerk's contact details on the council's website or by calling the switchboard. The mayor (or convenor in Scotland), usually acting on the advice of the chief executive of the council, will usually decide if a deputation should be heard at a meeting of Full Council (see page 37). If it's about an issue that involves more than one individual or household, and it hasn't been heard before, then the deputation will almost certainly be allowed. If there are a lot of requests for deputations, then some may be postponed to the next council meeting.

Deputations are usually heard at the start of council meetings. With Full Council they are a good way of raising awareness among councillors. They are also likely to be heard by members of the public and the press because they tend to come to the start of council meetings. But do not expect to do more than raise awareness, as public council meetings rarely take decisions. You will get an official response from the council, but it will usually be a gentle murmuring of understanding about your concern.

If you want to influence a decision, then you need to lobby the relevant councillors on the Executive or in the Cabinet a long way ahead of the meeting that is making the decision. Do not wait until the relevant meeting because that will be too late – most decisions are taken in private in advance, not on the night in public. If you have a friendly councillor, then it is worth getting hold of the Forward Plan all councils are required by law to publish. That will tell you what reports are coming up for discussion a long time before they appear on public agendas.

Questions to Executive or Cabinet Members

This is a good way to ask difficult questions and I find that a lot of helpful information makes its way into the public domain in this manner. Only councillors can ask questions of Executive or Cabinet Members, but once you've found a friendly councillor, he or she should be willing to help you.

Motions

Motions passed in Full Council do not technically bind the Executive or Cabinet, but if a majority of councillors vote for the motion, and it is worded in such a way as to call on the council to do something, then it makes it hard not to do it. For example, in 2009 Bradford's opposition parties proposed and passed a motion calling on the council to sign up to the 10:10 Campaign. The minority Tory administration abstained but they nevertheless felt obliged to sign up in the fullness of time.

At the end of 2008 Nottingham City Council passed the UK's first peak oil motion, which was partly about raising awareness and partly about providing a trigger for action.

> "The Peak Oil Motion sought to increase awareness amongst elected members about the debate around depleting natural resources. Cllr Graham Chapman [Deputy Leader] and I wanted to make our colleagues and staff of the council aware that we do not only need to think urgently about our commitment to reducing the effects of climate change but that we have the added pressure that oil reserves are falling and that global production of oil may have peaked.
>
> As a result of the Peak Oil Motion we commissioned the drafting of our City's energy strategy,

which contains all the commitments we/business need to invest in (renewables, reduction in usage and increased insulation) in order to make ourselves more resilient as a city and to commit to a low-carbon future. We are already the most energy-resilient city in the country – we create 4 per cent of the city's energy locally – but we need to do much better. The depletion of oil reserves is likely to lead to increased volatility in oil prices, and therefore other fuel prices – all of which have a direct impact on how we fund services within the council. Fuel price increases mean less money to spend on direct services, so it is very much in our interest to isolate as much of our council from that as possible. That means weaning ourselves off the oil habit."

Cllr Katrina Bull, Executive Member for Environment on Nottingham City Council

In 2008 Somerset became the first council in the UK to pass a motion in support of the Transition movement. Few elected members in Somerset understood the implications of it at the time. It was put forward by a Liberal Democrat councillor, Paul Buchanan, who had been involved with the setting up of various Transition Initiatives in Devon and Somerset. (See Chapter 3, page 59, for more on the Somerset motion.)

Leicestershire County Council passed its own Transition Motion in December 2008. Following on from that, North West Leicestershire District Council passed a motion recognising peak oil and the Transition movement. In September 2009 Bath and North East Somerset became the latest UK council to pass a motion supporting the principles and ethos of the Transition movement. Their motion did not create a budget for Transition groups, or seek to align the council's budgets with Transition thinking, but sought to set up a principle whereby Transition teams in the area could take their suggestions and requests to the council and get a sympathetic hearing.

It's worth noting that Nottingham's Peak Oil Motion was driven by just two people – Cllrs Bull and Chapman. Leicestershire and North West Leicestershire's Peak Oil and Transition motions appear to have been the background work of only one person –

The Nottingham Peak Oil Motion

This Council acknowledges the forthcoming impact of peak oil. The Council therefore needs to respond, and help the citizens it serves respond, to the likelihood of shrinking oil supply but in a way which nevertheless maintains the City's prosperity. It acknowledges that actions taken to adapt to and mitigate against climate change also help us adapt to issues around peak oil.

It will do this by:

- developing an understanding of the impact of peak oil on the local economy and the local community
- encouraging a move across the city towards sustainable transport, cycling and walking throughout the city
- pursuing a rigorous energy efficiency and conservation programme through its carbon management plan and leading on raising energy awareness across all sectors to reduce dependency on oil-based energy in the city
- supporting research and production within the city which helps develop local effective alternative energy supplies and energy-saving products in order to encourage a move away from oil-based fuels and also in order to create local 'green collar jobs'
- coordinating policy and action on reducing our city's carbon dependency and in response to the need to mitigate and adapt to climate change and peak oil.

Neville Stork, a former councillor on the county council and a climate change officer at Leicester City Council. Somerset's Transition Motion was also put together by just one person – then Deputy Leader Cllr Paul Buchanan. These people are what I call prime movers (see page 50).

Motions like this and the prime movers who create them are incredibly inspiring, but you do need officers and other councillors to be prepared to pick up the ball afterwards. Leadership and inspiration can come from one person or a small group, but substantive action can come about only if the wider institution is on board.

Power

Power in a council runs down two parallel hierarchies – the political and the operational. The politicians set strategy and the officers carry out the day-to-day running of the council. At the top of the political hierarchy is the Leader or elected Mayor, then come the Deputy Leader, Executive or Cabinet Members, Opposition Party Leaders, Scrutiny Committee Chairs, Champions (if they exist), administration backbenchers and finally opposition backbenchers. Officers are led by the Chief Executive, who manages Directors, Assistant Directors, Heads of Service, senior officers and junior officers. You can influence both branches of the system, but you have to be aware that they have different motivations and so will respond to different approaches.

> "The theory is that politicians do strategy and officers do the management but the reality is that both are interested in both so it's a bit more blurred than that. It's inevitable really – that's what happens when you get to the top. Lots of politicians have had experience in management and so are tempted to get involved. Most officers think that's bad. But it's inevitable."
>
> Cllr Keith Moffitt, Leader of Camden Council 2006-10

What does the jargon mean?

There's an awful lot of jargon in local government. To quote from the Totnes Energy Descent Action Plan, which was written by Jacqi Hodgson, a Green Party councillor on the town council: "The plans and procedures used in our present bureaucracy have created a highly sophisticated system and the language used matches this. The need for plain English and support to understand the system for all involved, from those within the system to communities and citizens seeking to be involved in decision-making that affects their work, is clear in public consultation exercises. These frequently lead to a lot of frustration and a sense of 'us and them' from both local authority and citizen perspectives."[2]

That sentiment was repeated by many in responses to the online survey conducted for this book. "Too much red tape and local authority speak" was the constant refrain. Nevertheless, there are a few key acronyms that it's worth getting your head around.

Local Strategic Partnership (LSP)

Local Strategic Partnerships are supposed to bring together representatives from all the strategic partners in an area – the council, the emergency services, universities, business and the voluntary sector – to address local problems, allocate funding and discuss strategies and initiatives. They aim to encourage joint working and community involvement, and prevent 'silo working' (i.e. different agencies that share aims working in isolation), with the general aim of ensuring that resources are better allocated at a local level. Local authorities are the lead partners. In Scotland, equivalent partnerships are called Community Planning Partnerships and in Wales Local Service Boards.

It's well worth trying to get a place on your Local Strategic Partnership or the equivalent where you live. It's a great place to spread the word about your agenda and to influence key players in your area. Most LSPs will have thematic working groups, which is where the interesting stuff gets done. Bristol's Peak Oil Report was commissioned by the city's LSP. Camden's Transition groups were set to be given a place on the borough's LSP in autumn 2010.

The voluntary sector is represented on the LSP, and Transition Initiatives and other environmental groups are part of the voluntary sector, so they have every right to be there. Do an internet search for 'local strategic partnership' and the name of your council. You should be able to find a contact name at the council for LSP matters so you can approach him or her directly.

However, probably the best route to get on to the LSP is to make contact with the existing voluntary-sector representatives. LSP meeting minutes are public, so you can find the name of the voluntary sector representatives and get in touch. They probably feel most marginalised on the LSP and therefore are likely to genuinely appreciate the support.

> "As Chair of Wandsworth Environment Forum, I am on the LSP and provide a link between Transition Initiatives in the borough and council decision-making processes."
>
> Denny Gray, Wandsworth Environment Forum

> "We've done a lot of work with Lambeth and they've been very supportive, e.g. given us meeting rooms and opportunities to speak at council events, but we haven't really been very organised about our ask. For example, we haven't yet requested a seat on the LSP, which we probably should do."
>
> Duncan Law, Transition Brixton, London

Local Area Agreement (LAA)

The LAA is an agreement between members of the LSP on which matters of mutual interest they wish to work on together. Progress against stated goals was monitored by the Audit Commission for public authorities; however, in 2010 it was announced that the organisation was to be abolished and its work to be done instead by the voluntary or private sector. Ideally you want to see your community group's aims reflected in the LAA, which means getting a place on the LSP.

Local Area Forum (LAF) or Local Area Committee (LAC)

Many councils have now introduced a decentralised level of decision-making, usually at the ward level, which are sometimes called Local Area Forums (LAFs) or Local Area Committees (LACs). They function a bit like parish councils and so are more often found in urban areas than in rural areas. They're chaired by councillors and can discuss local concerns. They can require council officers to give presentations, answer questions or simply to listen. Councils are supposed to take into account the views of LAFs as a reflection of opinion in a particular area.

In practice it's often only when LAFs are given budgets that councils start to pay attention. Camden's Local

Area Forums (LAFs) were each given a budget of £10,000 in 2009-10. In the Royal Borough of Kensington and Chelsea they get £100,000 a year each! Camden Council initially allowed only a small number of possibilities for how the money could be spent. I challenged that, because I wanted our LAF to be able to spend the money on food growing. We won on the point of principle and now a number of LAFs in Camden have chosen to spend their budgets on food growing.

I think Local Area Forums will become more important in the future, but to start with there'll be a constant battle between the council and the community over the limits of action and rules of engagement. Either way, it's a good place for community groups to get involved.

Sustainable Community Strategy (SCS)

Section 4 of the Local Government Act 2000 (England and Wales) says that every principal local authority must have a community strategy "for promoting or improving the economic, social and environmental well-being of their areas, and contributing to the achievement of sustainable development in the UK." The Local Government and Public Involvement in Health Act 2007 builds on that by requiring a long list of partner organisations to cooperate with the local authority in developing the Sustainable Community Strategy for the area, and in setting local improvement targets. The intention of these laws is that local authorities should work with the voluntary sector and private sector, as well as local people, to agree the content of the SCS. This sounds great in principle, but in practice it can be a bit meaningless – lots of worthy platitudes and no clear route map. Still, as we'll see in Chapter 9 (Procurement), the more that environmental goals are explicitly mentioned in the community strategy the easier it is to argue eco-priorities in procurement.

Total Place

'Total Place' is a whole-area approach to public services.[3] Under the scheme, which was piloted in 13 areas of England in 2008/09, local spending from across all local agencies is looked at as a whole, to focus on designing services around the needs of the customer and cutting out waste and duplication.

In March 2010 the Treasury announced that the 'Total Place' programme would be rolled out to the whole of England and that it would give "new freedoms from central performance and financial controls, to collaborate with strategic partners, to invest in prevention, e.g. of crime and anti-social behaviour, and to drive growth." By implication, action on carbon and sustainability is also something that should be covered by Total Place.

CRC Energy Efficiency Scheme

The Carbon Reduction Commitment (CRC) Energy Efficiency Scheme is a UK mandatory cap-and-trade scheme for large organisations in the private and public sectors.[4] The CRC will apply to organisations that have a metered electricity consumption of more than 6,000MWh per year, but not to those that are large enough to fall within the scope of the European Union Emissions Trading Scheme. A total emissions cap will be set by the government. Emissions allowances will be auctioned and can then be traded. Most large councils are in the CRC Energy Efficiency Scheme. The government has suggested that all local authorities will be obliged to join in due course.

Local Carbon Frameworks

In January 2010 the Department of Communities and Local Government approved nine pilot programmes

– Local Carbon Frameworks – aimed at exploring ways to reduce carbon emissions from homes, businesses and transport.

Manchester, Leeds, Bristol, Nottingham, Plymouth, Oxford, Haringey (in London), Northumberland and three councils in Dorset are part of the £3m pilot, which is designed to encourage local authorities to lead the way in setting clear targets for progress in meeting reduction targets, including strategies and delivery plans, as well as identifying support that could best be provided by central government.[5]

National Performance Indicators

In April 2008 the government published a new National Performance Indicator set[6] with the aim of reducing the number of targets imposed on councils and making area-wide working the responsibility of the LSP.

Until it was scrapped in 2010, the Audit Commission judged how well local authorities were performing against the National Indicator set, some of which involve working with strategic partners in the area. Poor-performing councils were named and shamed, and central government could intervene if necessary. For the first time councils and their local strategic partners were given sensible environmental targets. Previously the only environmental indicators were things like questions to residents along the lines of 'Do you like your bin green or brown?'!

Key environmental National Performance Indicators

NI 185 – CO_2 reduction from local authority operations
This indicator covers direct and indirect emissions from council operations. Direct emissions result from equipment and vehicles owned and controlled by the council. Indirect emissions result from services that the council has procured. The indicator includes schools but not social housing.

NI 186 – Per-person reduction in CO_2 emissions in the local authority area
This indicator is intended to measure emissions that can be influenced by local authorities and their partners, and therefore excludes direct carbon emissions from installations in the EU Emissions Trading Scheme, and road transport emissions on motorways. Emissions from power generation and refining have been included on an end-user basis. The Department of Energy and Climate Change (DECC) outsources the collecting of NI 186 data to a third party. This is by far the most challenging environmental indicator.

NI 187 – Tackling fuel poverty
This indicator measures the percentage of people receiving income-based benefits and living in homes with a low energy efficiency rating.

NI 188 – Planning to adapt to climate change
This indicator measures how well councils are doing on adaptation to climate change. All local authorities must report their progress against this indicator, which aims to embed the management of climate risks and opportunities across all levels of services, plans and estates. It's a process indicator that gauges progress of a local area in:

- assessing the risks and opportunities comprehensively across the area
- taking action in any identified priority areas
- developing an adaptation strategy and action plan

setting out the risk assessment, where the priority areas are (where necessary in consultation and exhibiting leadership of local partners), what action is being taken to address these, and how risks will be continually assessed and monitored in the future

- implementing, assessing and monitoring the actions on an ongoing basis.

NI 189 – Flood and coastal erosion risk management
This indicator measures the progress of local authorities in delivering agreed actions to implement long-term flood and coastal erosion risk management (FCERM) plans; these plans are Shoreline Management Plans (SMPs) and Catchment Flood Management Plans (CFMPs).

NI 191 – Residual household waste per household
This indicator measures the amount of waste that is sent to landfill, incineration or energy recovery divided by the number of households in an authority area.

NI 192 – Percentage of household waste sent for reuse, recycling and composting
This indicator measures the percentage of household waste sent by a local authority for reuse, recycling, composting or anaerobic digestion. The government expects local authorities to maximise the percentage of waste reused, recycled and composted.

NI 193 – Percentage of municipal waste landfilled
This indicator measures the proportion of municipal waste landfilled. The government's strategy on waste is to move waste management up the 'waste hierarchy' and divert an increasing proportion of waste away from landfill.

NI 194 – Air quality
This indicator measure the percentage reduction in noxious emissions (NOx) and particulates emissions (PM10) achieved through changes in a local authority's buildings and operations.

NI 197 – Improved local biodiversity
This indicator measures the proportion of local sites where positive conservation management has been or is being implemented. All local sites in a local authority area are assessed, not just those owned by councils. Evidence that the site is being enhanced rather than just being maintained is required and is used as a proxy for positive biodiversity outcome.

NI 198 – Children travelling to school
This indicator measures how children travel to school, i.e. the mode of transport usually used.

But it's not just on environmental indicators where community groups can work together with councils; there's a range of community indicators that fit perfectly with what the Transition movement in particular is trying to achieve:

NI 1 – Percentage of people who believe people from different backgrounds get on well together in their local area.

NI 4 – Percentage of people who feel they can influence decisions in their locality.

NI 6 – Numbers participating in regular volunteering.

NI 119 – Self-reported measure of people's overall health and well-being.

NI 151 – Overall employment rate.

These are all things that community groups can help with. The more that community groups can align their actions with National Indicators to contribute to positive progress on the performance of councils, the more likely they are to get a favourable hearing from local authorities (see page 50).

It's worth pointing out that the Local Area Agreement is simply a collection of National Performance Indicators that a particular area has decided to prioritise. As a Transition Initiative or community group, once you know which performance indicators are in your LAA you'll be in a good position to offer your help.

How can you get involved with emergency or resilience planning?

Civil contingencies planning (formerly called emergency planning) entails preparing responses to major emergencies such as severe weather, natural disasters, industrial accidents, transport accidents and terrorism. Under the Civil Contingencies Act (2004), councils have a duty to provide and maintain Emergency Response Plans so that staff know what to do in the event of a foreseeable emergency, whilst allowing council services to continue to provide as far as possible their usual day-to-day services.

Since 11 September 2001 London has had a Resilience Partnership. In the early days it focused its attention on preparing London for a terrorist attack. The London Mayor's Office says this played a vital role in the efficiency and speed of the emergency response during the bombings of 7 July 2005. In 2010 the partnership was preparing for how London would respond to pandemic flu, large-scale evacuation and severe weather conditions.

The London Resilience partnership includes the Government, Mayor of London, Metropolitan Police Service, City of London Police, British Transport Police, London Ambulance Service, London Fire Brigade, Association of Train Operating Companies, City of London, Environment Agency, Greater London Authority, Health Protection Agency, London's Business Community, London Councils, London Coroners, London District Military, London's Faith Communities, London Local Authorities, London Resilience Team, National Health Service, Network Rail, Port of London Authority, Transport for London, London's Utility Companies and London's Voluntary Organisations. So pretty much everybody really.

There's a resilience website – www.londonprepared.gov.uk/londonsplans – including a mass evacuation plan, which is interesting in the context of a peak oil crash. It also has a voluntary sector resilience forum, designed to improve the effectiveness of the voluntary sector contribution to emergency responses.

Most other areas of the UK have followed London's lead on this, so try doing an internet search using the

Resilience and faith groups

In one North London borough, Catholic churches have developed a network of emergency contacts based around the Deanery. There is a good working relationship with the police and the council, which was tested during the London bombings on 7 July 2005. Following this, the three Roman Catholic dioceses in London, in collaboration with Barking and Dagenham Primary Care Trust, have been working together to develop a means of mobilising emergency responders, coordinated by local deans, when needed.[7]

keywords 'resilience', 'emergency planning', 'civil contingencies' and the name of your area. There's also a government resilience website: www.cabinetoffice.gov.uk/ukresilience/preparedness/ccact.aspx.

The founder of the Transition movement, Rob Hopkins, and one of the most influential writers on peak oil, Richard Heinberg, have had a number of public discussions about whether Transition groups should be preparing for disaster or engaging in emergency planning.[8] To put it crudely, Richard is in favour of disaster planning because he thinks we are about to hit the wall, whilst Rob is against it because he is concerned about Transition becoming marginalised as just another survivalist cult.

We had the same discussion in Transition Belsize – should we have a subgroup looking at what we could do if supply lines break and civil order deteriorates? Well of course that is exactly what civil contingencies teams think about. So it seems to me that there are two possible responses: 1) start planning from the bottom up but without making it the raison d'être of a Transition Initiative; and 2) try to get representation on your local civil contingencies team and potentially also on the regional resilience forum.

What legislation might be useful to community groups?

There are some key pieces of government legislation with which community groups should familiarise themselves if they want to lobby effectively for change.

Local Government Act 2000

The 'well-being powers' in the Local Government Act 2000 (England and Wales) was used by Nottinghamshire County Council to set up a company to guarantee the supply of wood-fired burners and wood chip, because they wanted to move all the schools from coal to biomass. Essex used the Act to set up a municipal local bank (see box, opposite).

The equivalent power to advance well-being came in Scotland in the 2003 Local Government Act. The Act gives the local authority the power to do anything that it considers is likely to promote or improve the well-being of the area and/or people within that area, as long as it is not prohibited by existing legislation. Local authorities have considerable discretion in how they exercise this power. Scottish Executive guidance on the Act suggests a number of ways in which the power of well-being may be used, including: to promote sustainable development; to tackle climate change; to improve mental, social and physical health; to tackle poverty and deprivation; to promote financial inclusion in disadvantaged communities; to reduce inequalities and promote equalities; to promote local culture and heritage; to protect, enhance and promote biodiversity; and to improve community safety.

Planning Policy Statement 1 (PPS1) on Climate Change

Dover District Council used this to create an evidence-based renewables policy that went well beyond what was required by government (see Chapter 8, page 128).

Local Government and Public Involvement in Health Act 2007

There are decentralisation powers buried deep in the Local Government and Public Involvement Health Act 2007, which describe the duty of upper-tier authorities to divest budget and resources and assets to lower-tier authorities, including parishes and community

Municipal banks

In the wake of the 2008-09 financial crisis, four councils – Birmingham, Ceredigion, Essex and Liverpool – proposed setting up their own banks because high-street banks were failing to lend to credit-worthy local companies and home buyers. But only Essex actually opened a bank.

The council launched its Banking on Essex scheme, a joint venture with Abbey National (now Santander), in April 2009.[9] The suite of loans, intended to provide an initial £30m in new credit to small businesses, was portrayed as "one of the most significant developments in the governance or financing of Britain in 100 years".

However, by January 2010 Banking on Essex had lent just £263,000 to eight businesses. The council said that the scheme had been hampered by back office reorganisations at Abbey National resulting from its integration into the Santander banking group.

Birmingham ran the last UK municipal bank (from 1916 to 1976), but councillors apparently concluded that start-up costs for a new bank would be prohibitive and that lending might not begin in earnest until the recession was over.

groups, if a case can be made that decentralised assets can be managed better. So, for example a community group might feel it can manage a park better than a local authority and can start a petition of the local population to secure a change in council policy.

Under the petitions duty of the Local Democracy, Economic Development and Construction Act 2009 the council must set a threshold number of signatures for triggering a Full Council debate: no higher than five per cent of local residents using the population estimates published by the Office for National Statistics.[10]

Sustainable Communities Act 2008

This was designed to allow communities to suggest, via their councils, blockages that could be unlocked at the central government level. Communities loved it. Some councils embraced it. Some felt they were simply being used as conduits and refused to share ownership of their communities' ideas. Other councils simply ignored it and didn't bother to tell their communities about it.

But the real problem came in early 2010 after communities had sent hundreds of excellent suggestions to Whitehall via their local councils. One insider told me that no clear process had been designed to go from reception of those suggestions to action. Hopefully somebody will eventually unblock the legislation that was designed to unblock central government!

> "When our local council asked various stakeholders in the community to submit proposals under the Sustainable Communities Act, we were the only organisation that responded meaningfully. Therefore there seems to be a need for more public awareness about the need for Transition, and more effort from the council to engage the public meaningfully. Part of that would be for the council to provide more resources to engage the public, both directly and indirectly via the Transition groups."
>
> Philip Snow, Transition Newcastle-under-Lyme

How can community groups influence councils?

Where are the pressure points in councils? Honestly, they're almost everywhere. This agenda is so vast that every department and almost every council officer can do something positive to combat climate change, prepare for life after cheap oil and help to create a positive vision of the future. The area that seemed to me to offer the fewest opportunities was Licensing, but even there the council has to examine issues such as the recycling of glass bottles and food waste, or the use of energy by drinking and catering establishments. Scrutiny committees offer endless opportunities. Health Scrutiny and Children, Schools and Families Scrutiny are both places I would have gone if I'd had the time.

Here's a checklist for those seeking to influence council policy.

Striking gold – Rob Hopkins and the 2009/10 Mayor of Lambeth, Cllr Christopher Wellbelove, at the launch of the Brixton Pound. Photograph: Transition Brixton

Create the political will to effect real change

You can start by raising awareness among politicians. Being a local councillor is a pretty thankless task. Most councillors are completely overworked and are harassed by complaining residents left, right and centre. Councillors have feelings too. They like to feel appreciated. They love people who appear to bring solutions rather than problems; especially solutions that make them look good to their electorates.

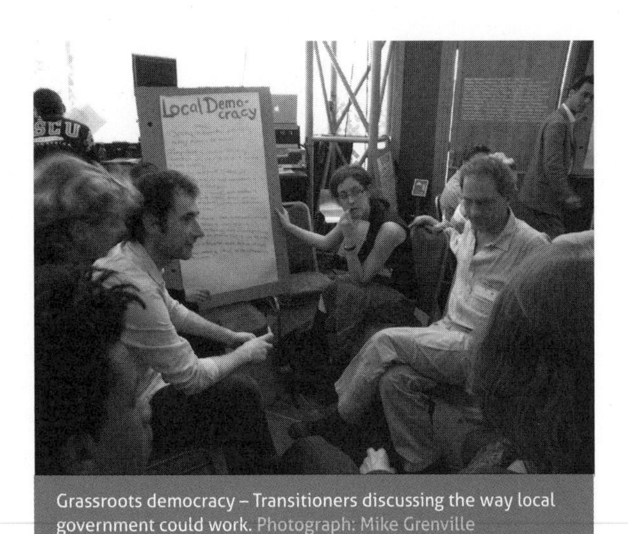

Grassroots democracy – Transitioners discussing the way local government could work. Photograph: Mike Grenville

Invite them to say a few words at the start of a film screening as a guest of honour. Ask them to cut the cake at your Transition Initiative Unleashing. Organise events in their field of vision, ones that are talked about in the community that elects them. Offer them first choice of dates when you plan an Open Space day. In an election period they'll probably be content to be Keynote Listeners, who take part in an event rather than speak *to* an event, as the local and parliamentary candidates in Totnes were in the winter of 2009/10. Above all, try to make them look good and feel appreciated.

Look for opportunities to create a one-to-one relationship with councillors, especially Executive Members.

A few key rules on approaching your council

- If you're representing a Transition Initiative or an environmental group, be careful not to blow it by going to the council too early. Make sure you've been through your organisational birthing pains and you know what you're asking for or what you're offering.
- Ideally you would have a physical manifestation of your project to show before you contact your council – proof that you're more than just another talking shop.
- Some groups start at the top and approach the Chief Executive or Council Leader. Others prefer to start at the bottom by asking for help with a project such as food growing. It doesn't really matter where you start – the main thing is you know what you're asking for or what you're offering to help with.
- When you do get your big chance, do dress smartly, don't take your family and the dog, and above all don't show too many slides with peak oil graphs on them. Don't tell council officers or councillors how unsatisfactory they are. Do be patient – remember that councils are big organisations.
- Above all, offer solutions not problems. And remember that councillors and council officers are human beings too!

The latter sometimes have relatively informal single portfolio public meetings, although I'm told that Camden is something of a rarity in this respect. These may be a good opportunity to start talking to the portfolio holder – before, after or sometimes during. You can also turn up at ward or advice surgeries and are unlikely to be turned away, especially if one of you actually lives in the ward.

The start and finish of Local Area Forums is another route to a conversation with councillors. Once you've engaged with them as allies rather than adversaries, ask them to sponsor motions in Full Council. Or ask questions to Executive/Cabinet Members. Don't be afraid to call your local councillors and invite them to have lunch or coffee with you.

> "The workload and entrenched or party-political views or lack of interest of many councillors seems to prevent them from getting actively engaged, or even attending events, seminars, etc."
>
> Susana Piohtee, Transition Hereford

> "I think the only thing I can say is that it is better to target individual councillors first to get one or two interested and involved. Hopefully they will then get a motion passed by the council to sign up to the Transition agenda and a few more councillors may then get involved. That is what happened when one councillor made the proposal that Abingdon should become a Fair Trade Council."
>
> A former Abingdon Town and Oxfordshire County councillor (quote sent to me anonymously)

> "We are working on raising awareness amongst our various councils. I have representatives from county, district, parish and community councils on the team, so we are in a good position. I recommend any new group to get their local councillors on board early."
>
> Will Hitchcock, Transition Nayland, Suffolk

> "Local councils are important and they do reflect the position of the town. It is important to make sure that they are accountable and that

there are people who are not on the council in attendance at as many meetings as possible. This keeps councillors on their toes and also shows them that the people are interested and do care about what is going on. After all, most local councillors are doing it voluntarily and it is a major commitment if they are giving it their best."

Andrew Capel, Energy Descent Llanidloes, Wales

Tailor your arguments

Different arguments will work with different people. Finance directors will listen if you can come up with ways to save money, maybe by cutting energy bills or using volunteers. Disaster planning officers will listen if you can point to possible breakdowns in supply lines or food riots. Many council officers will be motivated by morality concerns and a genuine desire to improve local communities, even if their statutory duties and financial pressures may blunt their ability to act in this regard. Officers may be particularly willing to engage with Transition movements in councils where quality of life surveys are undertaken regularly and regarded as a significant way of measuring the council's performance. Ways to make people happier might be a compelling argument in this respect.

Of course, a key argument would be simply to get on with it; to create physical manifestations of the project; to prove that the community can make this stuff work.

"We don't have to wait for politicians to tell us what to do, we just have to get motivated into taking action. If, as communities we can transform our own systems of living so that they are vibrant, upbeat, and meet challenges head-on with compassion and goodwill, then we can seed a wave of cultural diversity and social justice that will ripple across the world."

Sophy Banks, co-head of Transition Training

Align your goals with council indicators

Every argument you can make that aligns your goals with those of the council is a sure-fire winner. BioRegional, an environmental NGO, managed to persuade the London Borough of Sutton to build One Planet Living (see Chapter 1, page 30) into its core Sustainable Community Strategy. The 10:10 Campaign, which in 2010 is seeking to persuade individuals, businesses and organisations to cut their emissions by 10 per cent, uses National Indicator 185 (see Chapter 2, page 43) as the basic methodology for councils wishing to sign up.

Chief Executives are under pressure to deliver against central targets, and budgets are always being squeezed. Anything you can do to help is likely to go down well. Which would you rather hear: "We're heading for disaster and your policies are all wrong" or "Doing ABC can help you make progress vis-a-vis XYZ targets and we can help to get community support, and look here's an example of how it works."?

Look for a prime mover to drive change or a friendly insider to help open doors

Cllr Paul Buchanan was clearly the prime mover in Somerset, the person who kickstarted change. But the prime mover doesn't need to be a politician. Woking's Finance Director (later Chief Executive) Ray Morgan understood the benefits of investing in energy efficiency, combined heat and power (CHP) and renewable generation in terms of saving him money in the medium to long term. He was the prime mover in Woking.

Outsiders can also be prime movers. For example, Jeanette Orrey, the dinner lady at St Peters Primary School in East Bridgford, Nottinghamshire, did a huge amount to pump-prime the healthy school food movement, which Jamie Oliver then mass-marketed.

And it's probably fair to say that the reason Lambeth Council is getting to grips with the climate change and peak oil agenda is at least partly because of the motivational powers of Duncan Law, the coordinator of Transition Town Brixton.

TRANSITION SCILLY AND THE AONB OFFICE

Jonathan Smith, Transition Scilly

We formed Transition Scilly in late 2007 and have had strong connections with the Area of Natural Beauty (AONB) office from the start. One of our core group, Clare Lewis, works there part-time as an Environmental Awareness Coordinator, and I have had personal connections with the AONB office for a while.

The Isle of Scilly AONB office is there to look after the unique landscape, and its remit covers every issue from social interactions with the landscape to beach cleaning, environmental awareness and advising the Planning Dept on development issues. Essentially it is the first port of call in the council on any environmental issues.

The current AONB lead officer, Trevor Kirk, completely understands the concepts of peak oil and climate change, and all the issues that spin out from that. The AONB's latest five-year Management Plan has, for the first time, given very serious consideration to climate change issues. That's largely thanks to Trevor (with some prompting from me!).

Also closely related to the AONB's work is that of the Climate Change Coordinator, David Senior. He also thoroughly understands the Transition concept and we've had some very productive meetings with him. He's been pushing his job remit far wider than simply mitigation and adaptation to climate change, thinking much more about the whole range of community responses required.

One of our group, Nick Lishman, sits (with a Transition Scilly hat on) on the Joint Advisory Committee that manages the AONB – so has an input into the direction of its work. This also shows that we are taken seriously by local politicians, the council, landowners and other organisations. Nick also sits on a committee that decides the fate of project funding under the £2.2m Local Area Agreements programme (essentially European Union money). This has enabled us to enact some really positive input to the grant delivery process.

Photograph: Jonathan Smith

In essence we're working towards these goals with our council:

- Short term – gaining awareness of resource depletion and how that affects our community.
- Short term – helping to guide the work of the Climate Change Coordinator.
- Medium term – getting the council to adopt the principles of Transition (like Somerset did).
- Longer term – ensuring that the council is fully engaged (and taking a lead) in the process of community carbon cutting and resilience building.

I think the reason why we have such a good relationship with our council is because Scilly is so small (population approximately 2,000), so it's easy to know exactly who you should be speaking to and just make those connections. A lot of 'business' here is done on street corners and generally out of offices!

Even buildings can be prime movers! The success of BioRegional's BedZED low-carbon community in Hackbridge in the London Borough of Sutton was probably the main reason why the local council eventually adopted BioRegional's One Planet Living principles as part of its key strategy and then decided to make Hackbridge a low-carbon community.

If you really want to make progress, then find a sympathiser as high up in the council as possible and ask him or her to help with an organogram of those with power and influence. Those people should be your targets. Befriend them if you can. This is the approach I would take if I were trying to sell something to a large organisation, and that's basically what you're doing.

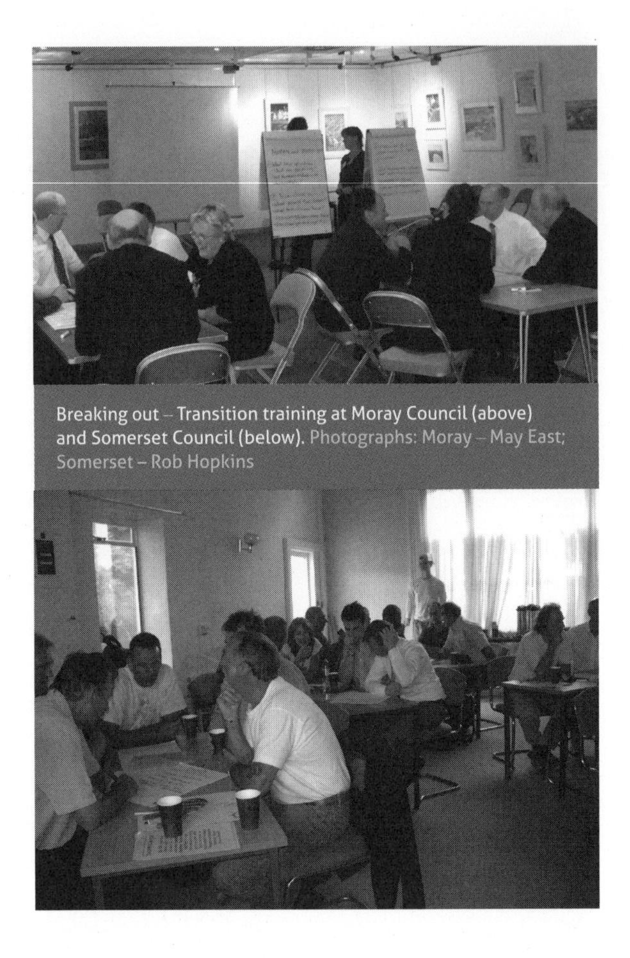

Breaking out – Transition training at Moray Council (above) and Somerset Council (below). Photographs: Moray – May East; Somerset – Rob Hopkins

Give the council the in-house skills and structure to deliver the new agenda

In May 2006 Camden was blessed with an army of planning officers specialising in design and conservation, but nobody who knew anything very much about environmental issues. So whenever experts came to address the Sustainability Task Force I arranged for them to give seminars to our planners on green roofs, grey water recycling, Passivhaus, etc.

The Transition movement has begun organising workshops for council officers on peak oil and resilience (see box, opposite). In councils where the politicians are sceptical about climate change, a focus on training that looks at energy scarcity and energy security as well as enhancing community usually goes down well.

If you're on the inside, i.e. elected, then lobby for a new structure. It took me more than two years to get an Assistant Director for Sustainability in Camden. My ultimate goal was a Department for Ecology (minimisation of natural resources), Resilience (ability of the community to deal with external shocks) and Well-being (happiness), or maybe an Executive or Cabinet Member for Energy Descent and Transition. In the end the next administration in Camden opted to make its Deputy Leader the Cabinet Member for Sustainability. Not quite as forward-looking as my suggested remit perhaps, but I was delighted to see them placing so much importance on green issues.

Ideally a Director-level officer (reporting directly to the Chief Executive) should be responsible for sustainability, or at least an Assistant Director. Lower than that and they won't have enough political power to do much for you.

Set yourselves up as experts

Some parts of the council require specialist knowledge. This is particularly true of planning and procurement, which, coincidentally, are two of the biggest levers councils have. Community groups therefore need to build up expertise, because councils and councillors listen to experts. For example, Conservation Area Advisory Committees (CAACs) often play an excellent role at planning meetings in helping councillors to ask the right questions or to understand the key issues.

Don't give up – do find a support group

Many of the respondents to the online survey conducted for this book said that the biggest obstacle was time and people.

Transition training for local authorities

In 2009 the Transition Network started piloting Transition training for councils, starting with pilots at Moray in Scotland, Somerset and Norfolk. Lead Trainer Naresh Giangrande set out the following four aims.

1. Engagement with the central ideas around Transition and particularly peak oil and resilience.
2. Challenging 'business as usual' through scenario planning and looking backwards from the future to correct the inbuilt bias in the system, which thinks that tomorrow will be much like today.
3. Exploring ways of working between councils and Transition Initiatives in their area in a mutually beneficial way – win-win.
4. Visioning a relocalised positive future – understanding that many things that strengthen local communities are National Performance Indicator deliverables (see page 43).

I helped out with the final pilot at Norfolk County Council, where a total of 80 council officers took part. Naresh dealt with peak oil and challenging business as usual. I talked about what a Transition council might look like and some of the ways that Transition groups can help councils to achieve their goals and performance indicators. They all seemed to appreciate it, so my sense is that there's an appetite for more Transition training among councils.

Ahead of her time – Doris Day teaches Transition Training in the 1958 film Teacher's Pet. Image: Ben Brangwyn with thanks to MGM!

> "There's so much to do and so little time. Therefore training up others to be able to engage effectively is really important."
>
> Denny Gray, Transition Wandsworth, London

> "The biggest obstacle is having enough people to back up those who are capable and relatively fearless about meeting and speaking at meetings. It wears people out and unless there is new blood and subs on the bench, your influence fizzles."
>
> Scott Morrison, Friends of the Earth Bath

> "I find myself doing it all on my own. No one else has the time to read all the stuff and attend the meetings in working hours. And I'm constantly sidelined by the active core of officers who desperately need to meet deadlines and can't wait for volunteers. I'm an alien in the local authority milieu."
>
> Michael Dunwell, Transition Forest of Dean, Gloucestershire

I know how Michael Dunwell feels. It takes a lot of energy and commitment to keep coming back for more when the answer is no, no and no again. But remember that all significant change movements have been rebuffed at the first, second and third times of asking. You have to keep asking. And keep improving your knowledge base. And keep building your coalition for change. And get yourself a support group, because this can be a gruelling business. What I hope this book shows is that even if it seems as though you are struggling to make headway, whether you're a councillor, an officer or a resident, every bit does count in the end.

> "Whatever you do will be insignificant, but it is very important that you do it."
>
> Mahatma Gandhi

Be patient – councils are big bureaucracies

I often use the analogy of a tanker in terms of the time it takes to turn things round on a council. They are big bureaucratic organisations with enormous rule books. Compare that to Transition groups who, at least in their early days, often test the limits of 'bounded chaos'. To the untrained eye, or even the trained eye, it's not necessarily an obvious fit. Some Transitioners see councils as fortresses and describe their interactions with it in terms of war metaphors.

> "We tend to think of local government as this fortress that we have to try and invade, make deputations to, and apply battering-ram techniques to. Yet the people in the fortress probably believe they'd like to do more and the public won't let them."
>
> Angela Ruffle, Sustainable Redland and Transition Bristol

> "We must find a chink in their armour. We must find the right words to use on the right people. We must also understand that some councils are highly conservative (note small 'c') and regard Transition with suspicion. To them we are pie-in-the-sky, woolly headed liberals and dangerous radicals. We must overcome this cynicism. We must be seen as being respectable, authoritative and as conservative as them. I would like to see pre-packaged 'events' that we can deliver to councils in a language they will understand, sourced not from Friends of the Earth or Greenpeace – no, this needs to be seen to come

from NASA and Whitehall. If we can convince the council that they will be held responsible for achieving a low-carbon economy and that it is inevitable AND the ONLY way of achieving it is through an Energy Descent Action Plan built with full community involvement, then we have a case. We need to be that bright and intelligent with some of these councils. If not, then we will be seen as a group of nutcases."

Mark Brown, Transition Town High Wycombe, Buckinghamshire

Others see patience as being the key skill Transitioners and environmental campaigners need.

> "Local government needs time to understand the aims of Transition groups and how the two can work together. Your council does not have all the answers and endless amounts of funding, but it needs help from the community or voluntary sector and will support Transition groups when it can see that they are trying to achieve similar goals. It may even fund projects once trust is allowed to build up."

Chris Rowland, Transition Town Lewes, East Sussex

WAVERLEY BOROUGH COUNCIL AND TRANSITION HASLEMERE

Clive Davidson, Transition Haslemere

The initiator of the Transition project in Haslemere contacted Waverley Borough Council when she was trying to set up Transition Haslemere. She met with the Sustainability Officer in September 2008 to establish what the council's strategies were regarding climate change and also how the group and the council could work together.

The council's Sustainability Officer had heard of Transition through Transition Farnham, which is also within the Waverley area, and offered help and support by whatever means possible. Waverley has limited funds and resources to carry out this kind of work in the communities and really saw the value in supporting the community to do it themselves.

Over the next 12 months Transition Haslemere received several hundred pounds for specific purchases from the council, the loan of ten energy monitors, printing of numerous posters and leaflets, lots of helpful information and press support. The Sustainability Officer also participated in our first Open Space meeting.

In October 2009 the three Transition Initiatives (Farnham, Godalming and Haslemere) held an Open Space meeting with Waverley at their offices, where we worked out how we could help each other and what our aims and plans are. This effectively started our Waverley Transition Hub, which meets regularly.

In November 2009 the three Transition Initiatives met with Surrey County Council's Climate Change Officer, again to introduce ourselves and our aims and see how we could work with Surrey.

At a town level we have had a local town councillor on board since the beginning (operating in a personal capacity). This has proved a useful introduction to the town councillors as well, some of whom attended our Open Space meeting. They have been very supportive and provide us with free use of the town hall for all of our meetings.

The town council sees us as helping to fill a much-needed role and is happy to support us. For example, the Haslemere Initiative, an organisation set up to support initiatives for the local economy, environment and community, has written our Transition project into its action plan, as it was weaker on the environmental part than others.

"I think the thing is to be patient and persistent. People need to recognise that councils and councillors are very constrained by established procedures – unlike Transition, which is a grassroots organisation that can wake up in the morning and decide whatever it wants to do. So in some ways that's uncomfortable for a council, because first of all as a Transition Initiative we don't have a very defined structure and someone in charge and all those kind of things, and that's kind of scary because they're used to dealing with things that are very structured."

Mike Grenville, Transition Forest Row, East Sussex

"The council is a hierarchical governmental institution with its own agendas and processes. Residents, including ourselves, have a very wide range of views about what (and how) the council does or doesn't do, or indeed should and shouldn't do. Hence we have had to ensure that we are totally independent whilst prepared to work in partnership with them on specific projects/issues where practicable. It's important to be independent, patient and strong!"

Dave Morris, Sustainable Haringey, London

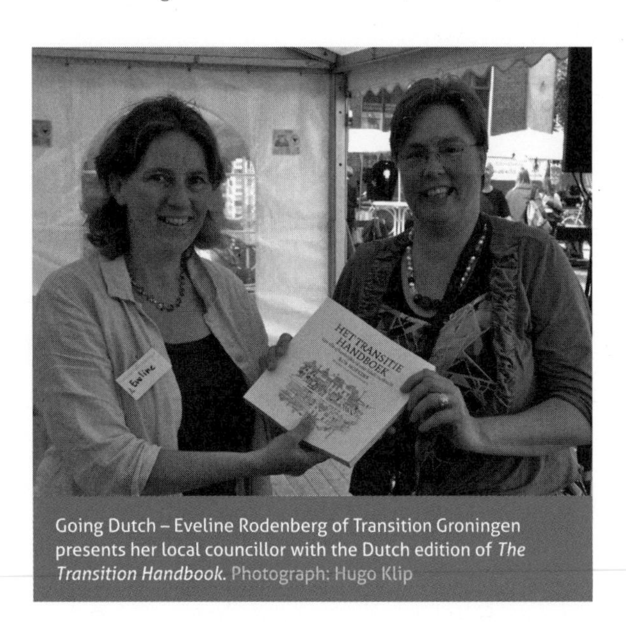

Going Dutch – Eveline Rodenberg of Transition Groningen presents her local councillor with the Dutch edition of *The Transition Handbook*. Photograph: Hugo Klip

Try to reach your council through different community groups

Some of Camden's schools were on to the environmental agenda far earlier than the council and set up a Sustainable Schools Forum, which the council belatedly joined. The NHS is now starting to be a force for good in sustainability terms, notably on the need for us all to eat less meat (see Chapter 7). Some businesses are also forging ahead on the carbon agenda. There is potentially a powerful coalition out there. It's worth getting hold of the Local Strategic Partnership agendas to see who they go to and mapping that against the local players who are doing good work on the sustainability agenda. Those who figure in both columns should be your allies.

Almost anything can be changed if enough pressure is brought to bear, but the pressure needs to come from many different voices.

Try to engage or befriend council officers

In the previous chapter I advised community groups not to be in too much of a hurry to make contact with councils but to wait until they had a project that they wanted help with or there was a process they wanted to try to influence. When you have a project, approach a particular officer. Just ring up the switchboard and keep asking until you find someone. These days the fact that many councils have climate change/carbon/ sustainability teams makes it much easier to know where to start.

> "We realised it was all about finding the right person who would make stuff happen – who in the end was an officer and not a councillor. We had sent deputations to talk to Scrutiny Committees – and nothing came of it. It was going 'outside of the machine' and just having a meeting with someone that can get things going. It is the officers that actually do the work and have the expertise, so they are worth making friends with."
>
> Chrissie Godfrey, Taunton Transition Town, Somerset

> "The experience of a local councillor and a local government officer are entirely different, particularly in terms of what they can deliver. This needs to be highlighted in a positive way. It is also entirely possible for a local community group like a Transition Initiative to have a good working relationship with one sector of the council on food or energy but not other sectors. This can be the case especially where the political side of a council hasn't signed up to peak oil or isn't working with the Transition movement."
>
> Louise Shrubsole, Sevenoaks Council, Kent

That's sage advice from Louise, who was an officer at Sevenoaks Council when this book was written. Others have managed to get themselves jobs as officers as a result of their work in Transition and other community eco-projects. In truth there is no one single way in. Different groups are doing it different ways. Some choose the political route; some the officer route. Some remain on the outside and build bridges; others use battering rams. As I've said previously, if I were selling a product to someone, or trying to get a story as a journalist, then I would use all possible approaches.

In 2009 the Community Development Foundation published a very handy short guide on how to work with your council on the environment and sustainability.[11] Its key recommendations are as follows.

- To be successful, your community or voluntary group must be persistent, positive and ready to work with others.

- You will need patience and networking skills to find the right person within the council who is interested, either formally or informally, in sustainability issues and can make things happen for you. Also, you have to bear in mind that the

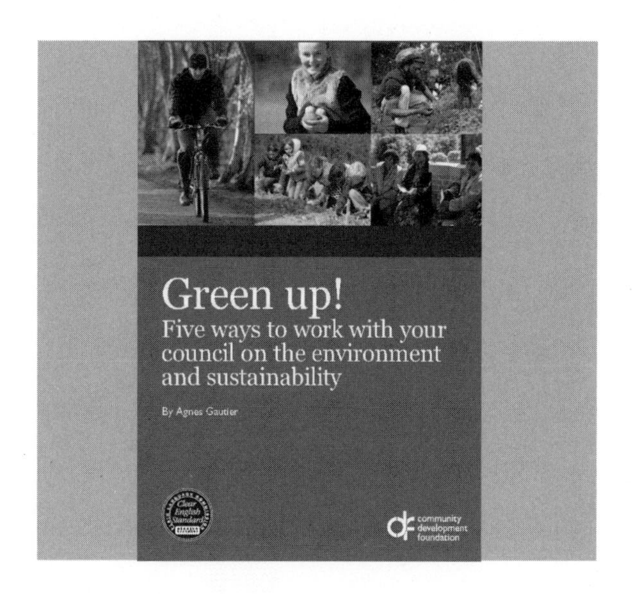

council's administrative procedures might take longer than you expect.

- You must be ready to bring something to the table. Council officers must be able to see what your group can contribute to their targets.

- Often the council will be dealing with third-sector representatives already. You will often have to go through these people first to reach out to the council. You will also make more of an impact if you approach your council with a coalition of many similar organisations.

That makes good sense to me. Engaging with your local council can be hard work but the rewards can be worth it, and for some groups it's been an extremely positive experience.

> "We have been very lucky. So far we have been given everything that we have asked for. We haven't approached town planning yet, I anticipate a little resistance when we go there!"
>
> Maggie John, Transition Town Hervey Bay, Queensland, Australia

> "Our biggest success has probably been gaining the ear of and support from our Chief Executive, Cabinet Member for Housing and Environment, and Head of Forward Planning, in theory at least. But the constraints they work under can prevent wholehearted engagement. We have a very good working relationship with Herefordshire Council's Sustainability Unit though. And we had an invitation to talk to City Council about Transition and Energy Descent Action Planning."
>
> Susana Piohtee, Transition Hereford

Chapter 3

THE VIEW FROM THE INSIDE – THE EXPERIENCES OF FOUR ECO-COUNCILLORS

Cllr Paul Buchanan, Somerset County Council

Paul Buchanan is an ideas man. Actually he's a whirlwind. He's clear about what can be done and is determined to do lots. He was elected to Somerset County Council in 2005 and, being a capable sort of chap, was given a wide-ranging portfolio containing the environment, biodiversity, sustainability, county farms and the economy. By 2006 he was Deputy Leader of Somerset Council and had won a South West Politician of the Year award.

In 2008, after a visit to the Transition mother ship in Totnes, Paul suggested to the Liberal Democrat group on Somerset Council that it put forward a motion to Full Council supporting the principles of the Transition movement. The Lib Dems liked it because it contained lots of localism and eco-activism, both of which have always been classic ingredients of Lib Dem politics. Paul then persuaded Transition groups to lobby their councillors, which is why, he says, the motion was passed unanimously and Somerset thereby became the first council in the UK to pass a motion in support of the Transition movement.

"The motion was very simple," recounted Paul two years later. "It was disarmingly simple. Because it was so simple nobody read it properly. The officers didn't check it. The Members didn't read it properly. The press were there. There were Transition people from all over the county. They'd all lobbied their local councillors. The councillors just thought it was a nice community motion."

The Somerset Transition Motion

This Council:

- Acknowledges the work done by communities in Somerset on Transition Towns and that the independence of the Transition Movement is key to its grassroots appeal;
- As demonstrated in its Climate Change Strategy, fully endorses the Transition Town Movement and subscribes to the principles and ethos of the organisation's goals to reduce dependence on fuel oil and create more sustainable communities;
- Commits to providing support and assistance to all towns in Somerset that wish to join this initiative to help them achieve the goals they set for themselves as local communities, as demonstrated under the 'Community Initiatives' section of the Climate Change Strategy;
- Therefore, requests the Scrutiny and Executive Committees to consider through the council's strategic planning process:
  - Allocating funds to assist in achieving the outcomes of the Transition Towns Movement in Somerset,
  - Requiring all directorates to engage with and provide support for Transition Initiatives in Somerset;
- Through the work outlined above, seeks to become the first Transition Authority in the UK;
- Agrees to undertake a review of its budgets and services to achieve a reduction in dependence on fuel oil and produce an energy descent action plan in line with the principles of the Transition Initiative.

One senior council officer told me: "It came as a complete surprise to us. This was not something we'd argued for. The first I knew of this was when the resolution was passed by Council. We didn't have a problem with it per se but there were issues for us about resourcing. The motion was passed. We were committed at a very senior level and given instructions to do all this stuff, but we didn't get any resources for it."

Somerset County Council agreed to do a mapping exercise of local authority activities in accordance with the aims and objectives of the Transition movement, and a training event for councillors, heads of service and directors. My contact said: "There was quite a lot of nervousness about the training. Knowing the Transition movement, we were a bit concerned that hugging might be involved, but it went very well actually."

Challenges of the Transition Motion

However, it quickly became clear that the Transition Motion had opened a bigger can of worms than just resource pressure and the threat of spontaneous hugging, as the officer I spoke to made clear. "We were suddenly being asked to do all this stuff to fulfil the objectives of the Transition movement, but they're not an elected body, and we serve an elected body, so there were difficulties there for us as officers. Although Transition groups have been very successful in getting together people who are concerned about climate change, etc., they do not represent the whole community, nor can they claim to, whereas the local authority is elected by the whole community, so there's a difference there for us, and for me as a local government officer."

This is an important point. It's quite true that Transition initiatives and other special-interest groups do not represent entire communities. Even if many people do not vote and are turned off by conventional politics, local elections do give residents a chance to vote for their representatives. So, for the time being, councillors represent communities much better than Transition groups do, even if in some cases the representation is more theoretical than actual. That's one reason why it helps to have councillors on board in some shape or form, and why in the longer term it may help to get eco-activists elected (see Chapter 14).

The other problem is one that most of us have come across in our dealings with the Transition movement – the concept of 'bounded chaos'! Paul Buchanan says: "The Open Space concept – 'whoever turns up is the right person' – is great if you've got an external framework and constraints, but if you don't, then you can create anarchy. I've been through this process half a dozen times with different Transition groups. When you get together with a Transition group you can't just map out a hierarchical structure, you have to let them get going, it has to be about community engagement, so you sit back and let them evolve and try to nudge them towards decisions and actions. Most of them have absolutely no experience in structure, meetings, outcomes, process or even the confidence to make a decision. Quite often they're too nice to take decisions. Nobody wants to lay down rules, everybody's subscribing to all the good stuff in Transition, but it can be a real block."

According to my contact, officers at Somerset Council were quite simply overwhelmed by Transition groups. "We were getting requests saying 'Look, you as an authority have signed up to these principles and objectives, and therefore you should do this, this and this.' And we were getting emails from all these people in the Transition movement saying 'What about doing this? And supporting that? Blah blah.' For which? Why? There's no resource for it. Bearing in mind that the first we knew about this was when the motion was passed, it does beg the question as to whether the authority knew what it was letting itself in for."

Things eventually settled down and agreement was reached on a way forward. "We've gone through the mapping exercise, we've provided them with a lot of information, and I think we've helped them to get organised in dealing with us," said my source at Somerset Council. "We've said 'Look, we need to know who it is we're dealing with, particularly about training where we're trying to organise a meeting with the top tier of Somerset council, we need to know what your programme is, what your objectives are, what you are going to be delivering, what you are going to get out of it, what is it going to take, and who's going to be there and how many people at a time, etc. etc.' We had to say that to them because they would just turn up as a loose structure and everybody would go home happy. If we're talking about volunteers, that's fine, but if it's Members and Senior Officers, you can't carry on like that."

This is, I think, a salutary lesson for all community groups. Be clear about what you want and why you want it before you approach an organisation as big and as structured as a council. There is a real risk for Transition groups that a lot of goodwill can be lost in the early days by trying to do too much, too chaotically, without widespread support. My advice is to engage on small projects when you know what you want but not to try to engage across the whole of the council until you have a proper structure.

I became aware of Somerset when Rob Hopkins contacted me to ask if I could help with Somerset Council's vision of what a Transition council should look like. Within hours I was also contacted by a consultant who been drafted in by Somerset Council to help them understand what a Transition council should look like. Nobody had a clue and nor did I. It was the blind leading the blind.

The Executive Summary of the consultant's report is contained in full in the Appendix of this book. It's an important document because it seeks to point up transferable lessons from Somerset. That's utterly critical, because Somerset's period as a Transition council may end up being a short one – in May 2009 the Conservatives won control of the council and by mid-2010 it was still unclear whether they would significantly progress the Transition council concept.

Here are the key lessons from the consultant's report, which still hadn't been published in full by the time this book went to print.

- Staff were positive about the motion and the review.
- Some departments have policies that are contrary to the aims and principles of Transition, e.g. Housing & Development, Food & Farming, Land & the Environment.
- There is wide inconsistency in assessing the twin challenges of climate change and peak oil across services and directorates.
- There has been no contingency planning for steep fuel prices and/or scarcity.
- More training would be beneficial, as would a Transition Officer.
- There is potential conflict between Transition ideals and the need for Somerset to do business in the real world.

The man who started it all off – Cllr Paul Buchanan of Somerset – is an Icarus figure. He made enemies by rising too fast. He was, in my opinion unfairly, reported to the Standards Board by Somerset's Chief Executive and then shabbily deselected by his own Lib Dem group. But even if Paul flew too close to the sun, and the concept of Somerset as a Transition council is one that may not outlive the county's current political leadership, there's enough of lasting value from the Somerset experience for others to be able to build on.

In the words of Henry David Thoreau: "Though I do not believe that a plant will spring up where no seed has been, I have great faith in a seed. Convince me that you have a seed there, and I am prepared to expect wonders."[1] Or, as the Spanish proverb goes: "More grows in the garden than the gardener sows." It turned out that the seeds sown by Cllr Paul Buchanan at Somerset County Council bore fruit just down the road at Taunton Deane Borough Council.

A resilient Taunton Deane

Over three months in the summer of 2009, over 350 staff and 26 councillors from Taunton Deane Borough Council attended 11 half-day workshops, led by volunteers from Taunton Transition Town, with the aim of creating a vision for the future of the area. Plumbers, planners, environmental health officers and car park attendants mixed with senior strategy officers, carpenters, elected Members and tree surgeons. In and of itself, this was described as extraordinary by participants because on no other occasion, apparently, have the people who make up this organisation come together in this way.

From that rich mix of backgrounds, skills, interests and political leanings a story emerged – a surprisingly consistent story about what a resilient Taunton Deane might be like in 2026. Together the group envisioned a future in which the borough has developed a successful low-carbon, 'localised' culture and infrastructure, capable of withstanding upheavals in the wider world likely to occur as a result of our changing climate and increasingly expensive and insecure oil and gas supplies. Topics that were explored included homes, water, food, shopping, transport, leisure and holidays, work, education and training. There was, without question, according to those who were there, a shared sense of urgency that things need to change, and change soon. People at the workshops were, for example, prepared to use the car for fewer trips, or to car-share, and to walk and cycle more, and insulate their homes more effectively. They were prepared to shop differently, prioritising locally produced food and goods and buying seasonally.

There was also a growing sense that Taunton needed to get back in touch with what 'community' is – though less clarity on how to achieve this. People called for a far stronger lead from the council itself. They wanted to see a commitment to actually making a difference, with councillors taking the politics out of carbon, able to be brave enough to be honest about the facts: that change has to happen now.

One of the first actions the council took was to announce that all 18,000 planned new homes would be carbon neutral, with the authority saying 'no thanks' to developers who refuse to conform to these new standards. From that point on, all new builds would have to use sustainable building materials, with compulsory energy efficiency measures.

There were many calls for both more time and more space to grow food, with suggestions that the council create more allotments, with a greater choice of smaller plots. However while the attendees recognised the council's role in promoting best practice, no one expected the council to 'do it all'. Most importantly, attendees said that they were ready to talk to their family and friends about the issues, to get across just how important changes on an individual level could be.

Towards a resilient Taunton Deane, the output from that summer in Taunton, is listed in the Resources section. It's the phoenix that rose from the ashes of what Paul Buchanan started at Somerset County Council.

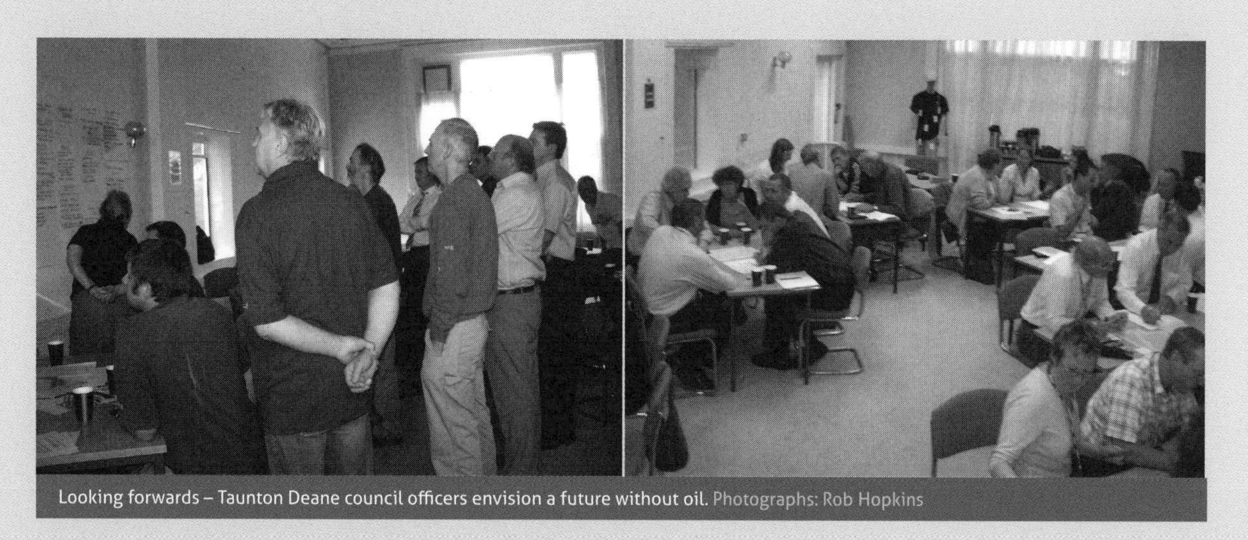

Looking forwards – Taunton Deane council officers envision a future without oil. Photographs: Rob Hopkins

Cllr Fi Macmillan, Stroud District Council

Fi Macmillan was elected to Stroud District Council in Gloucestershire as a Green Party councillor in 2008. But what's really interesting about Fi is that she became a councillor because of her involvement with Transition Stroud. A Green Party councillor who was also in Transition Stroud suggested to Fi that she become a council candidate. Fi accepted and was duly elected in May 2008. In other words, she went into politics because of her concern about climate change and peak oil, and because of her involvement in Transition.

"The other Green Party councillors told me there were a lot of opportunities to influence even in opposition," Fi told me in late 2009. "I felt I could better support those officers trying to do good stuff if I were a councillor. I also feel that I've added credibility because I'm a councillor. I sit on planning committee and make sensible decisions – I give Transition Stroud more credibility. I'm a bridge to local government. I don't think I use this enough yet. I'd like to move this forward on more substantial projects."

Transition Stroud developed an informal relationship with Stroud District Council, involving regular meetings to swap information on progress. The Local Strategic Partnership (LSP) set up a Global Changes Think Tank, to consider the twin impacts of climate change and peak oil. Fi became a leading light in the organisation but she found the experience quite frustrating.

"The Think Tank was being used to inform the Local Development Framework [LDF – see Chapter 8, page 131], which I think is OK although at the time I felt Transition Stroud wasn't being given enough credit. But it was frustrating because without the political will to make the changes we identified there wasn't any change. The LSP Food Strategy was a case in point. Transition Stroud was leading the process and we apparently had total support for the concept of Local Food, but no partners came forward to actually work on it.

"We had a food conference in November and all of that food stuff is being used as evidence for our core strategy [in the LDF], but there are no new people getting involved doing new food stuff. Maybe more minds are open as a result. I don't know. The council has contracted out its Green Transport Week – Trip Switch – to Transition Stroud, so that's good.

"I've achieved some small things on planning committee on renewables, etc., but I can count on one hand the number of things I've got through on that basis. The more important work has been behind the scenes. For example, I gather we won the government Pay-As-You-Save bid [see Chapter 5, page 92] partly because of the close joint working between Transition Stroud and Council.

"There's a lot of statutory stuff we have to do and, although I work hard at it, it's not the reason I was motivated to stand. It doesn't speak to my deeper values on sustainability. I found it really challenging to move into the chamber and see how different my agenda was from the other councillors'. However, after listening to what was important to them I concluded that there's a shared agenda which I might not have seen before.

"If I talk about climate change, they switch off. They don't get it or believe it. But if I talk about resource depletion and energy security, they're interested. They want to see us produce more energy locally and I think we can make an economic case for it. So I've changed my whole dialogue from sustainability to energy. But I feel like it's taken 18 months to see the landscape clearly and what's possible. I've learnt the language now but it took a long time.

"The jury's out for me as to whether this a good use of my time. I'm going to spend the next two years researching ESCo [Energy Service Company – see Chapter 6, page 107] models and see what we can achieve. I think the whole renewables/energy agenda can work in tandem with a mainstream issue like fuel poverty. This appeals cross-party. I want to develop an ESCo that cuts carbon emissions and protects tenants from fuel poverty at the same time. But if I can't achieve anything through the council, then I'll go to the private sector or the not-for-profit sector.

"It's hard. I'm a single parent. I pick up my kids from school. I see my colleague writing posts on his blog every day and communicating to his electors, which I'd like to do but I just haven't got time. That said, I am recommending to others that they stand for the council. We need more women in politics and we need more younger people because they see life in a different way. But we also need more support for those who work and have children. At the moment it's all geared toward those who are retired."

Cllr Linda Hull, Glastonbury Town Council

Rising up out of the wetlands of central Somerset, Glastonbury is perhaps most famous for the festival of the same name and its Tor, the highest point for miles around. Interestingly, despite the town's progressive image, Glastonbury Town Council is a bastion of traditional values. Many members of the council struggle with what the media and popular culture think Glastonbury is about. So when Linda Hull stood for election, she knew it would be a challenge.

"Fourteen of us independents all sympathetic to Transition – under the banner of Glastonbury United – stood for 16 seats in May 2007. One got elected. There was a by-election six months later. I stood again and got elected. We didn't do any real electioneering apart from publicity in the papers. The second time round I knocked on doors, which made a considerable difference.

"Every month I report to the town council on the activities of Transition Glastonbury. My main success has been awareness-raising about the issues and the activities of our Transition group. This is mainly due to the fact that, as a town councillor, I've discovered that this layer of local government doesn't really have a lot of power or influence, in and of itself. Some basic physical changes have happened in the town hall, itself, such as new energy-efficient lighting and loft insulation. I do try to ask awkward questions and ask my fellow councillors to consider the bigger picture; I try to keep stating the rationale for increasing our resilience. The council has regularly supported the harvest show established by Transition Glastonbury food group by providing venue hire for free. We've been supported by the presence of various recent Conservative mayors at our events and we've been successful in winning some funds from the Local Strategic Partnership to develop some small projects.

"It's been a mixed experience to be honest, because of the kind of council it is. It's split down the middle – Tories versus Lib Dems. There are only three women. I'm the youngest councillor, at 41, with many already retired and some over 80. Glastonbury is quite a divided community, so if Transition gets associated with one party then the others will reject it.

"Standing as an independent has perhaps been one of the most difficult things. It's been quite isolating. Maybe I should have joined one of the main parties. My natural inclination is Liberal but the power really lies in the hand of the Conservatives here. And since many of the Tory councillors are very sceptical about the reality of a changing climate and resource depletion, that would have been a very difficult step for me to take. I think a lot of people feel glad I'm there but standing as an Independent – you have no party machine. It's hard work.

"I don't know if I'd stand again. That's mostly because of the isolation. I'd say either don't stand alone or if you're going to stand, then make sure you find some good personal support – your own Heart & Soul group, as it were. It can be quite grim sitting round a table like that when they don't even talk to you. It's not for the fainthearted, and as someone with no family history of political work, you have to learn on the job. It can also be demoralising to find out how powerless the lowest level of government is . . . but it's been an education. I remain convinced that it's local collective action that will make the difference, but mobilising that is the toughest challenge we face.

"One of the most interesting and rewarding things I've done is with the emergency planning committee. Basically Glastonbury is an island in a drained wet land. We're on a floodplain, so obviously we're prone to flooding. Emergency Planning brings together such a wide range of people – military types, music festival folks with dreadlocks who know about rigging electrical systems, street food sellers, trauma counsellors, the Women's Institute. I found it very enjoyable and empowering."

Alexis Rowell, London Borough of Camden

Before I was elected in May 2006 there was only limited political commitment to climate change and no understanding of peak oil in Camden Council. Camden had achieved Fair Trade status as a borough, we had the highest headline recycling rate in central London (see Chapter 10, page 154, for the details behind the headlines), we were a leader in car-free housing (see Chapter 11, page 170) and we were very well regarded on sustainable procurement (see Chapter 9, page 137, for how the wheels came off on that one), but that was about it.

The few Camden officers who were working on sustainability issues in 2006 were dispersed around the council and had no managerial direction because senior officers had zero incentive to do anything about climate change or carbon reduction.

That said, I wasn't the only one getting frustrated. Those who were on the Executive in Camden kept telling me that they also struggled to get relevant and timely information out of officers. So maybe it's endemic in the system.

I was quite often too far out in front of the other Task Force councillors – the problem of the locomotive and the carriages. I wanted to change things, to change a lot of things, and to do it fast. I sometimes made the error of not making sure that everyone in the Task Force was aware of what I was doing or where I was going. I think the same was true but in a different way in the Council Chamber. There was often a sense that people could leave the sustainability agenda and concentrate on other stuff because 'Alexis will sort it out'. That sometimes made me feel a bit isolated, although I know that most councillors appreciated what I was doing and the energy I was expending.

I was assigned Camden's Assistant Chief Executive, Philip Colligan, as my initial point of contact. I felt it was a great fit. He seemed to understand where I wanted to go and tried to prevent me from bouncing off the organisation. Together we worked out that the constitution gave nobody control over the sustainability agenda. Like the newly created Eco Champion, it was in a constitutional void. It was there for the taking.

Sadly, though, he went off to become Deputy Director of Housing and I was given a long-serving officer who was made Assistant Director of Public Realm and Sustainability. My new liaison officer quickly saw that the other members of the Executive, even the Lib Dem ones, were not particularly interested in the eco agenda. Or at least, to be fair to them, they were new councillors struggling with huge portfolios and didn't really have time to push sustainability. And the Conservative Environment portfolio holder, although not anti-sustainability, was more interested in other issues and highly sensitive to the idea that I might be running sustainability policy from the back benches. My liaison officer saw which way the wind was blowing and became less and less able to help me.

A glass half empty

I came close to giving up in those dark days. The pace of change in Camden was painfully slow. It was tiny steps in the right direction rather than bold leaps forward. But a large part of the problem was that I underestimated how long it takes to move an institution like a council – especially when you're a backbencher – and I wasn't seeing all the changes. Patience is a virtue when dealing with anyone, but especially when dealing with a big bureaucracy.

But behind the scenes things were moving, albeit slowly. A Sustainability Team was in formation. The Head of Sustainability was starting to gain dotted-line management relationships with officers elsewhere in the council working on sustainability issues who, up until then, had had no managerial or political direction on the eco agenda. Although the Chief Executive of the top council in the country never really 'got it' in any shape or form, the arrival of carbon-reduction indicators from central government helped to focus her mind.

Philip Colligan used to describe me as a 'glass half empty' sort of person. Looking back, I recognise that we did a lot. With more political will among councillors and senior officers we could have gone faster, but we did at least start the process of change. And towards the end of my 2006-10 term of office I was hugely comforted to hear from middle-ranking officers I bumped into that they really appreciated my efforts and the fact that I was always challenging senior officers. They said it made their job easier.

Senior officers struggled with constant revolution. They wanted a delivery plan set in concrete to work to. But the world doesn't work like that, especially not the world of climate change and peak oil. Things change all the time. And I wanted to be able to continually revise our plans, whereas officers found moving targets difficult to deal with. This was never satisfactorily resolved.

One of my biggest problems was that I set myself up as an expert on the outside. I was often challenging officers and Executive Members because I thought I knew what I was talking about. And if I got pushback I went looking for more information to bolster my case. That creates a real problem for an institution like a council, where councillors are supposed to set direction and strategy but not get involved in detail, and where officers are supposed to be the experts and deal with operational issues.

Passivhaus was a classic example. I spent a lot of time and energy trying to highlight the importance of the Passivhaus standard (see Chapter 5, page 94), but they kept putting up barriers. That only made me more determined. In the end I organised a Passivhaus conference for London's planners, building control officers, architects and housing associations – simply because I felt I was getting poor advice from Camden's officers and their external advisers. The conference was a huge success, and its successor the UK Passivhaus Conference is now a biannual event that I organise.

Slow steps to success

I also read our planning framework carefully and started using the sustainability parts more forcefully. So, for example, the section on resources said "we will require developers to conserve water". That word 'require' allowed us to get a green roof, rainwater harvesting, grey water recycling and other forms of sustainable urban drainage into all developments. Sometimes it's all in the wording!

For most of the period 2006-10 the Sustainability Task Force members constituted about half of Camden's planning committee (Development Control) and started asking eco-questions based largely on what they had heard in Task Force meetings. Officers quickly learnt that their reports needed to have a significant chunk in them about sustainability issues and were able to use the concerns of the committee to push developers to be greener than they otherwise would be. We built up a powerful combination of forces all pushing in the same direction.

More than 100 substantial Task Force recommendations were accepted during 2007/08, and they led to the creation of a sustainability team and formed the basis of the council's new sustainability delivery plan. The Task Force led borough-wide climate-change consultation for Camden Council's Community Strategy. We helped introduce emissions-based parking policy and incentives for electric cars.

The Task Force Report on Energy Efficiency suggested: introducing a Revolving Energy Fund to finance energy-saving investments; tri-generation systems on housing estates to provide heat, power and cooling to local neighbourhoods and businesses; moving the council to carbon accounting; retrofitting Victorian properties to be low-carbon exemplars; and reworking eco-grants to favour insulation rather than solar panels and wind turbines.

We initiated an audit of recycling to consider environmental impacts instead of just tonnage targets, which led to a move from a commingling collection system to greater separation. We helped amend the council's food procurement contracts to enhance sustainability aspects. I set up the Carbon Disclosure Project pilot to carbon-audit Camden's largest suppliers – the first such work done by the CDP with a public authority, now the CDP Public Procurement programme. I also initiated a food-waste-to-biomethane trial as vehicle fuel for council vehicles – another UK first.

The Task Force Report on Transport including proposals on: introducing a general 20mph speed limit; setting up a Camden Carbon Offset Fund; building a comprehensive cycling infrastructure; establishing a free bike scheme, as in Paris; and expanding car club provision. The Task Force Report on Food, Water, Biodiversity & Green Spaces including proposals on: encouraging food growing, especially on housing estates; reducing meat availability on Council menus; building more green roofs; mapping and potentially revitalising Camden's 'lost' urban rivers; and creating wildlife corridors.

I helped launch the Camden Eco House (July 2008), a council-owned Victorian property refurbished to reduce carbon emissions by 80 per cent and opened to the public as a low-carbon exemplar. The Task Force secured the installation of a green roof on the Grade II-listed Camden Town Hall. Task Force members on planning committee made climate change and fossil-fuel scarcity key factors in all planning decisions. We won the inclusion of the Passivhaus standard as an aspiration in Camden's new Local Development Framework. Task Force members played a critical role in getting Camden to join the 10:10 Campaign and sign up to the Friends of the Earth Get Serious Campaign, which is looking for CO_2 reductions of 40 per cent by 2020.

Taken in the round, it was a bruising experience because I had little experience of how councils work and the council had little experience of dealing with someone working outside the rules with so much energy as me. Although we moved forward, the constant challenge meant I lost a lot of political capital on the way. And it cost me a lot of emotional energy.

However, looking back, it feels as though real progress was made in terms of bringing green issues to the fore at Camden Council. All Camden's political groups went into the 2010 local election campaign calling for an Eco Portfolio on the Executive and a target of a 40 per cent reduction in carbon emissions by 2020. I wanted to see the Task Force merged with the Leader's Group on Sustainability and become a subgroup of the Local Strategic Partnership (LSP), so as to widen out the agenda across Camden and offer a Camden Green

Summit in July 2010, organised by the new administration. This seemed likely to happen. The Task Force was a fantastic ideas body, its all-party membership was critical, and it attracted a lot of support from residents, but I felt it needed to be put on a quasi-constitutional footing and reach out to the whole of Camden.

At the end of the day, I felt I'd done a huge amount and had succeeded in changing the direction of the tanker slightly. I understood better how the councils worked and knew where to put pressure or influence decision-making. But I felt I had done what I could at Camden in the Eco Champion role, so I didn't stand for re-election in May 2010.

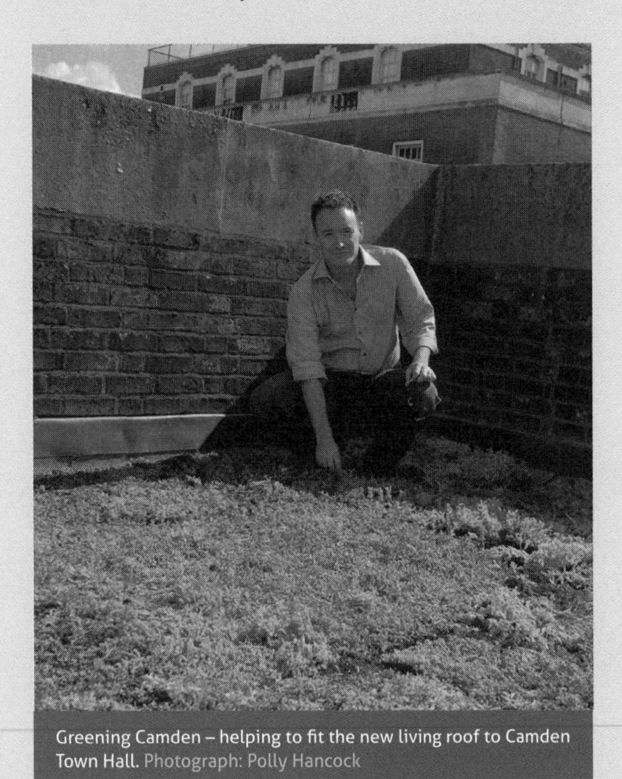

Greening Camden – helping to fit the new living roof to Camden Town Hall. Photograph: Polly Hancock

I have continued to work at the community level with Transition Belsize and Camden's Transition groups, using the incredible learning that being a councillor in Camden gave me, and at the national level with campaigns such as 10:10 and 'Old Home SuperHome' (see page 92). And I offered to help Camden's incoming Labour administration if they needed it: an offer they accepted with alacrity. Some things are far more important than party politics!

Chapter 4

BIODIVERSITY AND GREEN SPACES

Questions answered in this chapter:

- How can we enhance the perceived value of biodiversity in local councils?
- How can council estates be greened?
- How can we create more green spaces in cities?
- How much carbon can we lock up in nature?
- How can we measure biodiversity?

Biodiversity, the variety of life forms within a given ecosystem, is all too often forgotten in the global warming debate – which is why I've made it the first thematic chapter in this book! Worse, biodiversity is being steadily destroyed, mainly because the market-dominated view of the world is unable to price its value properly.

A lot of work has gone into pricing carbon, as part of the global effort to combat climate change, but very little of the excellent work being done to value forests, soils, wetlands, fisheries, species and coral reefs around the world has made it into the mainstream. In 2009 a mammoth three-year project funded by EU countries called 'The Economics of Ecosystems and Biodiversity' estimated that the continued loss of forests and biodiversity would cost us £1.2-2.8 trillion a year. "The economic invisibility of ecosystems and biodiversity is increased by our dominant economic model, which is consumption-led, production-driven, and GDP-measured," said study leader Pavan Sukhdev of Deutsche Bank. "This model is in need of significant reform. We are running down our natural capital stock without understanding the value of what we are losing."[1]

> "I think the majority of Londoners are completely divorced from nature. Bringing about a sustainable society has got to involve strengthening people's links with and knowledge of nature – whether that's through more protected green/open spaces and green roofs and enriching local diversity; growing more food in gardens, schools and allotments; more trees; or appreciation of water."
>
> Cllr Adrian Oliver, Camden Green Party and Sustainability Task Force member

One way to think of biodiversity loss is as an early warning system or canary in a mine. A 2°C increase in average global temperatures may seem superficially attractive to those of us living in northern climes, but for most plants and animals the speed of change is too much. We are asking them to adapt to changes that would usually take thousands of years, not decades. Scientists at the WWF have estimated that most species on this planet (including plants) will have to 'move' faster than 1,000 metres per year if they are to keep within the climate zone they need for survival. Climactic zones (isotherms) are moving towards the poles at the rate of 56km per decade, whereas plants and animals are travelling at only 6.4km per decade and are often confronted by insurmountable man-made obstacles.[2]

We are already seeing animal and plant species dying out at an alarming rate as the climate changes and man-made habitats restrict adaptation. In the mid 1990s Edward O. Wilson, the Harvard naturalist sometimes described as Darwin's natural heir, first estimated that about 30,000 species were becoming extinct each year. Further research has confirmed that just about every group of animals and plants – from mosses and ferns to palm trees, frogs and monkeys – is experiencing an unprecedented loss of diversity. The rapid loss of species we're seeing today is estimated by experts to be between 1,000 and 10,000 times higher than the 'background' or expected natural extinction rate.[3]

The 2007 assessment report by the Intergovernmental Panel on Climate Change[4] predicted that 20-30 per cent of all species would be pushed past the point of no return and condemned to extinction by 2050 if nothing was done before then to sharply reduce the burning of fossil fuels that cause climate change. The International Union for the Conservation of Nature (IUCN), the official keeper of the grim records of species loss, warned that "climate change will become one of the major drivers of species extinctions in the twenty-first century" and that we could be heading for a rerun of the Permian-Triassic Extinction Event 251 million years ago, when some 95 per cent of all species were lost.[5]

According to Mathew Frith of the London Wildlife Trust, we in the UK already know what mass extinction territory looks like. "Over 95 per cent of England's species-rich meadows have been lost since 1945 due to habitat loss and degradation, and the red squirrel is likely to be reduced to a few island refuges within 30 years if the trend in its decline continues."

How can we enhance the perceived value of biodiversity in local councils?

A community's role in protecting local biodiversity is effectively unique. In other words, while councils and communities can play only a small part in the much-needed global effort to limit global warming, if we do not step up our efforts to protect wildlife and habitation threatened locally then no one else will.

» *Enlist your local university or college to conduct research on council and community actions*

A vision of how an urban housing estate might look.
Photograph: Landlife

UCL and biodiversity

If you have a local higher education establishment, then it's worth seeing if you can work with it in some way. Masters courses are always looking for projects. Over the last few years Camden Council has worked with environmental masters courses at University College London (UCL) on a number of projects. One of my favourites was on biodiversity. I was concerned that biodiversity was being swept under the carpet at Camden Council. Not deliberately but, in the same way that markets fail to price natural capital, so councils fail to focus on issues, like biodiversity, which are a bit fuzzy.

MSc students at the UCL Environment Institute were given a remit to assess whether Camden officers understood their statutory duties with regard to biodiversity and to gauge the level of institutional commitment to action on biodiversity. They concluded that protection and enhancement of biodiversity fell between the gaps at Camden Council.

The vast majority of council employees didn't know they had statutory responsibilities for biodiversity and didn't see protection or enhancement of biodiversity as relevant to their work. Most departments said it was the responsibility of someone else. The majority of those who responded said there was no leadership on this issue in their departments.

The net result of the UCL project was a vastly improved Camden Council Biodiversity Action Plan, a new Biodiversity Manager and a clear line of management for biodiversity matters.

Communities and councils need to take a more holistic approach to all environmental objectives, and nowhere is this more true than in our approach to green spaces. These are key to so many environmental objectives, such as protection or enhancement of biodiversity, storm-water attenuation and carbon dioxide extraction – examples of what are increasingly known as 'ecosystem services'. Green spaces also add to our quality of life, although this is not yet well enough understood. 'Well-being' is receiving a lot of attention from academics and policymakers (see Chapter 13), but not yet in the context of the well-being benefits of the natural environment.

How can council estates be greened?

The biggest opportunity for 'greening' spaces, especially in urban areas, is often on housing estates. On these sites, and on council-owned land in general, the authorities have spent 50 years concreting and tarmacking over everything in sight – either to cover up contamination or because the resulting hard surface was low-maintenance. All too often the remaining green spaces are comprised of perennial ryegrass, which has little biodiversity value. According to Mathew Frith of the London Wildlife Trust, perennial ryegrass is also a variety developed for its durability and tolerance of what we throw at it, e.g. chemical pesticides and fertilisers, dog excrement, etc. The maintenance contract for Housing Department green spaces then becomes crucial. In the past this has been seen as a lowest-maintenance, lowest-cost affair – hence the ryegrass.

» *Lobby the Housing and Parks Departments via your local councillors to make sure maintenance contracts explicitly include reference to employing a biodiversity specialist to enhance green spaces*

Housing estates are also the ideal places to grow food. In many cases it doesn't make sense to dig up the concrete and tarmac, which in urban areas often covers contaminated land anyway, but rather to install raised beds and builders' bags, such as was done at the Vacant Lot food-growing project in Shoreditch in East London.

Vacant Lot – mini-allotments in builders' bags on an East London housing estate.

Since the roof and walls of every new development can potentially be green, that means the Planning Department has a key role to play in enhancing biodiversity. Equally, every building that can be retrofitted with a green roof should be. For example, according to a survey conducted for the Greater London Authority's London Plan and Environment Team,[6] about a third of roofs in the City of London have ballasted roofs to hold down insulation, which can be converted to green roofs immediately.

» *If you are a community group in an urban area, suggest a green roof audit to your council's Sustainability Team – it could work on greening them with the buildings' owners*

» *Lobby your ward councillors to ask Housing officers to conduct an audit of green spaces and potential green spaces on housing estates, and to set targets for reworking them to enhance biodiversity and other environmental objectives*

How can we create more green spaces in cities?

Another joined-up approach to enhancing biodiversity would be for Local Strategic Partnerships to work towards the establishment of more managed habitats and wildlife corridors. Camden's draft Local Development Framework says the council will:

- Protect designated open space and nature conservation sites.
- Identify parts of the borough that are deficient in open space and nature, and opportunities to provide additional open space and natural areas or improvements to existing spaces in terms of biodiversity, quality and access, including improving green links.
- Require developments to make a contribution towards open space and wildlife provision to tackle deficiency in the borough.
- Require the provision of new or enhanced habitat, where possible, including through biodiverse green or brown roofs and green walls.

Greening the City – a living roof on a building on the Greenwich peninsula in London. Photograph: Dusty Gedge

Totally green

Imagine an architecturally uninspiring council block on a run-down estate, with a central courtyard full of boarded-up shops and a lot of unpleasant and unused hard surfaces, which suffers from traffic noise and antisocial behaviour problems. What if the community and council worked together to install green walls, green roofs and green spaces, including food-growing areas, to try to achieve the following demonstrable changes?

- Reduced traffic noise pollution – improved quality of life for residents and reduction in mental health problems.
- Storm water attenuation – reduced risk of flash floods in the surrounding area as climate change brings more intense storms.
- CO_2 extraction by plants – reduced CO_2 per capita, a National Performance Indicator (see Chapter 2, page 43)
- Enhancement of biodiversity – another council target and a statutory responsibility.
- Reduced vehicle pollution – fewer health problems.
- Thermal insulation – reduced energy bills, fewer fossil fuels used, less CO_2 generated.
- Food growing – reduced food bills, reduced food miles, fewer fossil fuels used, less CO_2 generated.
- Reduction of the urban heat island effect – fewer elderly deaths, increased comfort for residents.
- Mitigation of wind in winter in the courtyard – increased comfort for residents, increased community cohesion through use of the proposed community garden.
- Cooling of internal ambient temperature in summer – fewer elderly deaths, increased comfort for residents and reduction of solar gain (shading).
- Improved aesthetics – improved social cohesion, reduction of antisocial behaviour, improved quality of life for residents.

Green walls are popping up all over the place now, so I'm sure it's only a matter of time before this totally green approach makes its way on to the drawing board.

- Ensure that new development does not adversely impact on biodiversity.

Brighton is going much further and is seeking to become the first urban biosphere city. Biosphere reserves are protected areas that are meant to demonstrate a balanced relationship between humans and nature.

> "It's a big challenge, and it could take 20 years, but I believe that we have got the ambition and the ideas in this city to step up to this challenge and make today's dream tomorrow's reality . . . We must look at ways to create green open spaces and corridors, grow local food, provide sustainable homes and transport, and encourage healthy lifestyles . . . The planning process is clearly key to any city's development . . . we have put in place a system that requires all new developments in the city to consider their environmental impact. Through the framework we have created we hope to see more buildings

embracing sustainability. Buildings with green roofs, green walls, solar panels and rainwater collectors . . . And buildings that are divided by great trunks of plant life that will act as the city's lungs."

Cllr Mary Mears, Leader, Brighton Council, October 2008

» *Lobby your council's planning dept to set a target for annual increases in green space in the Local Development Framework, and don't ignore the value of walls and roofs*

Green walls in London. Photographs: Biotecture

» *Engage with your local Wildlife Trust (www.wildlifetrusts.org) to help map your area in terms of actual and potential wildlife spaces and corridors, and then lobby for more*

Parks and open spaces

Of course, all cities already contain open spaces in the form of parks. Many of these are fabulous places that offer welcome respite from the stresses and strains of city life. However, there are often ways in which their natural resources could be better managed.

Councils waste thousands of pounds raking leaves out from under shrubs and trees in parks, thereby damaging soil habitat as well as removing soil cover, and often using leaf blowers – which generate carbon emissions as well as noxious and noise pollution. According to David Hughes, a retired professional gardener, "The general principles should be don't be too tidy, and avoid the temptation to keep prodding the soil with spade, fork or hoe. Insects don't mind plants in serried rows but bare, constantly tidied soil is a different matter." That's surely music to any permaculturist's ears!

Leaving dead wood where it falls is helpful to biodiversity. Unfortunately, tree surgeons and private gardeners are all too ready to cart the trunks of small trees away rather than making a design feature of them in gardens (either left in the ground or felled). Stag beetles may be iconic, but they need dead wood to survive, as do a great many species of invertebrates and fungi that help in the critical cycle of turning dead matter into soil. People also underestimate the practical and visual amenity value of the fungi associated with rotting timber. More work could be done to spread the word on this.

Councils have a statutory duty to control Japanese knotweed and giant hogweed on their land. The Transition Belsize Foraging Group started talking to the council about harvesting Japanese knotweed because it makes great jam (apparently) but, at the time of writing, officers felt that they had no choice but to continue to use pesticides to deal with these two plants. However, most less-troublesome weeds can be treated manually instead.

> "The majority of herbicide spraying takes place on the housing estates, as they have larger areas to spray and their sweeping specification isn't as comprehensive as in parks. Manual removal of weeds by hand costs four times as much as using herbicides, and it isn't as successful unfortunately.
>
> We can stop using herbicides if we invest more in the hard surfaces in our parks and on our housing estates, maintain a decent cleansing specification (sweeping) and accept that we will have increased maintenance costs from manual removing of weeds. Increased funding is the answer in reality."
>
> Peter Stewart, Parks Dept., LB Camden

. . . or we convert hard surfaces into green spaces, wild meadows even; we increase plant diversity so that nature can fight its own battles; and we change mindsets so that weeds are not seen as all bad.

» *Find out what your council's pesticide policy is and seek help from the Pesticide Action Network (www.pan-uk.org) to change it for the better*

Camley Street Natural Park

One of Britain's oldest and most influential urban ecology parks, and internationally renowned as a centre of excellence in environmental education, Camley Street Natural Park was created from an old coal yard back in 1984 through a unique collaboration between the London Wildlife Trust, the Greater London Council and Camden Council. Situated in the heart of King's Cross, the park features a valuable mosaic of habitats and supports a remarkable diversity of wildlife for its inner-city location. Over 300 higher plants have been recorded, including common broomrape (*Orobanche minor*), hairy buttercup (*Ranunculus sardous*) and common spotted-orchid (*Dactylorhiza fuchsii*). Breeding birds include the reed warbler, moorhen, wren and blackcap.

Managed as a community resource by London Wildlife Trust on behalf of Camden Council, the reserve has a visitor centre and is popular with all kinds of people seeking respite from the buzz of the city around them, as well as being a hub for London Wildlife Trust volunteers. It has a full-time education programme for Camden schools and is able to arrange activities for many other schools and community groups outside the borough. Camley Street Natural Park won a Green Flag Award in 2009.

An oasis in London – Camley Natural Park behind King's Cross in London. Photograph: London Wildlife Trust

How much carbon can we lock up in nature?

Another way to see biodiversity in the sense of trees, plants and soil is as a mechanism for extracting carbon dioxide from the atmosphere. Put simply, the more trees, scrub and grasslands there are, so long as they are not burnt, the more CO_2 they will lock up or sequestrate.

> "In order to lock up a lot of CO_2 you need plants that grow fast. Woody plants lock up a lot, although they grow more slowly. So when thinking about trees in parks, the best would be fast-growing, long-lived species. The fastest would be poplar, plane, willow, birch. But poplar and willow only have a 50-60 year lifespan. Beech or oak will take up less in the short term, but will store it for about 200 years.
>
> Grassland is generally felt to be quite good as grasses grow fast, and depending on circumstance some of this may be locked up in the soil. However, obviously lots of plant material is taken off park grassland when it is cut and this then decomposes, etc.
>
> We think that low-nitrogen fertiliser tends to mean that ryegrass grassland locks up more carbon than heavily fertilised grassland – so

parks planted to a fastish-growing grass that is not over-fertilised would do well.

It is also important not to disturb the soil too much (i.e. with cultivation and digging), as this tends to stimulate the release of the carbon that is already locked up in the soil."

Gareth Edwards-Jones, Professor of Agriculture and Land Use, Bangor University, Wales

Soil carbon and the Soil Association

In late 2009 the Soil Association published a lengthy report[7] arguing that actions to increase soil carbon levels have major potential for mitigating agricultural greenhouse gas emissions, and that organic farming practices lead to higher carbon sequestration than conventional approaches. They argued that organic farming increased soil carbon levels by producing additional sources of organic matter, creating organic matter in forms that are more effective at producing soil carbon, integrating crop and livestock systems, and increasing the proportion of vegetation cover that promotes the soil's micro-organisms that stabilise soil carbon. The report argues that soil carbon impacts of agriculture are ignored by current greenhouse gas accounting systems, which means that the current greenhouse gas emissions of agriculture have been greatly underestimated, the emissions of organic farming greatly overestimated, and the real potential of soil carbon sequestration overlooked.

The Soil Association suggested that widespread adoption of organic farming practices in the UK would offset at least 23 per cent of UK agriculture's current official greenhouse gas emissions. At a global level, the effects of agricultural soil carbon sequestration could be even greater: widespread organic farming could potentially sequester 1.5 billion tonnes of carbon per year, which would offset about 11 per cent of all man-made global greenhouse gas emissions for at least the next 20 years.

Set up an action group to talk to your local farmers about carbon and other environmental issues and see whether your council has a Farming or Sustainability officer who might get involved

Locking up carbon in trees

Trees lock up carbon, usually enhance biodiversity (except when they're on unimproved species-rich grasslands, heaths and downlands, where their presence can damage the biodiversity interest of these habitats), and have considerable amenity value, for example by providing shade on a hot summer's day. The London Borough of Richmond has an informal 'One out, two in' tree policy, but over the last 20 years the trend in most areas is for more trees to be cut down than are planted.

Trees on private land can be felled so long as planning officers do not regard the loss as detrimental to visual amenity. But this 'control' applies only to Conservation Areas, those trees protected by Tree Protection Orders and, outside the inner London boroughs, trees of a certain size (for which a Forestry Commission Felling Licence is required). Outside these constraints trees can be felled at will without any consideration of a tree's value in terms of carbon sequestration, biodiversity value and air quality.

I can think of two possible ways to deal with the problem of unnecessary tree felling. The first is simply to ban the cutting down of all trees in a borough, on public or private land, using a local authority's well-being powers (see Chapter 2, page 46). Clearly there would need to be exceptions, e.g. dangerous trees or trees causing significant structural damage or trees damaging important habitats. But the general rule could be that no tree is felled without permission and permission is hard to get.

The second approach might be to apply a price to a borough's trees – all of them – based on their value in terms of amenity, carbon extraction, air quality and biodiversity. This is an approach taken by Bonn City Council, where the starting price for removing a tree is €100,000 and more if the tree is deemed to be of particular value.

» *Lobby your ward councillors or your council's tree officer for a 'one out, two in' tree policy and/or a minimum price on the felling of a tree, which rises based on amenity and biodiversity value (see Chapter 7, page 122, for more on fruit and nut trees)*

How can we measure biodiversity?

One thing local groups can do that would be of huge value would be to find a way of measuring the biodiversity on a regular basis so that we are able to take stock of what we have and what is at risk. There are already residents' surveys of biodiversity, such as the London Wildlife Trust's stag beetle survey and RSPB's Big Garden Birdwatch[8] (for the past 30 years the RSPB has been asking its supporters to count the birds in their gardens), as well as borough-led garden surveys. These sorts of survey are incredibly valuable. They're the reason we know we've lost more than half of our house sparrows and three-quarters of our starlings.

URBAN TREES AND CLIMATE CHANGE

Vassili Papastavrou, biologist and a member of Bristol Tree Forum (council-led) and Bristol Street Trees (community-led)

With climate predictions for longer, hotter summers, the temperature in Britain's cities is set to rise, and more than one might expect.

The reason is the 'urban heat island' effect, which results in city temperatures some 3°C hotter than the surrounding countryside. When there are extreme temperature events the differential can increase to 10°C. Modern cities have many hard surfaces, which absorb heat and radiate it back.

'Urban green infrastructure' is the new buzz phrase. Put simply, it means vegetation, including grass and of course trees, and it keeps cities cool by providing shade and by evaporative cooling. The advantages of urban trees over other forms of greenery is that they do not take up as much space on the ground and can be fitted in to streets and parks, making our cities so much more pleasant to live in. And if the trees are deciduous trees that will become large, they provide the shade at just the times of year that it is needed, allowing sunlight to come through in winter.

Research by Manchester University[9] has shown that a 10 per cent decrease in tree cover can result in an increase in urban temperature of 7°C, increasing to over 8°C in town centres. Conversely, adding 10 per cent cover can keep our cities within the 1961-1990 baseline temperatures.

Yet all over Britain urban trees are being removed. A 2008 government study, Trees in Towns II,[10] shows that many are not replaced and those that are replanted are often ornamental species that will never become large – so-called 'lollipop trees'.

Bristol is no exception. We watched with horror as trees all over the city were being removed, with no genuine consultation and no opportunity to challenge the rationale for the chainsawing. For a while it seemed that Bristol was to become a city of stumps. One third of our

10,000 street trees were scheduled to be removed over a ten-year period. In order to challenge this problem we set up a small local group, Bristol Street Trees and through our work discovered that trees are being removed for four main reasons:

Bogus subsidence claims

The problem is not so much the trees but the inadequate foundations of our Victorian housing stock. In our area, the council responded to a vocal complainant and removed a beautiful oak tree against the wishes of the locals, in order to avoid the possibility that it might in the future just possibly cause subsidence to one building.

Concerns over risk

Around three people each year in the UK are killed by trees on public land, equating to a risk of one in 20 million. More people are struck by lightning. The risk of being killed in a motoring accident is around one in 17,000.

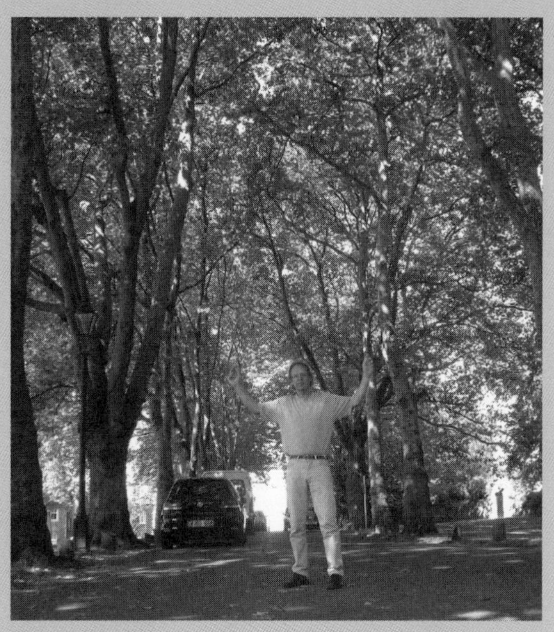

Fighting for urban trees – biologist Vassili Papastavrou.

Highways schemes

Here in Bristol, we have lost important trees to road revamps. Simply putting in an extra set of traffic lights on a junction resulted in the loss of several big trees.

Private householders

The loss of urban trees through applications by private householders is huge. Usually after the event the real reasons becomes apparent – whether it is building an extension or turning the front garden into a drive.

So what is the solution? We need to make it very much harder for urban trees to be removed. Road schemes should be built around existing trees, not through them. Householders should accept the benefits of living in tree-lined streets and stop worrying excessively about a few cracks – or move elsewhere. Concerns about the very small risks posed by trees need to be put in perspective. Trees with faults need to be managed in a way that allows their retention (maybe by removing offending limbs or decreasing the size of the canopy).

What we need is a moratorium on urban felling unless a tree can clearly be demonstrated to be dead, dying or dangerous. This is the key: if we cannot stop the chop, the loss of existing trees alone will make our cities pretty uncomfortable to live in. Not within a few years but right now.

And of course we need more trees planted. Flowering cherries, rowans and ginkgos won't help. Instead we need trees that will grow fast and become large – exactly the kinds of tree that councils often avoid planting now. Perhaps the best example of a tree that does well in the city is the London plane: it is extremely tolerant to the stress of city life and produces a magnificent canopy.

So, let's take a fresh look at urban trees, cool in so many ways.

See www.bristolstreettrees.org

Greenspace Information for Greater London (GiGL)

GiGL (www.gigl.org.uk) is the open space and biodiversity records centre for London. The survey results can be used to enable evidence-based decision-making by the London boroughs and conservation organisations. GiGL is partnership-led and holds data on London's open spaces, habitats and species (it has 1.3 million species records), and is increasingly being used by council planning departments on development control matters. GiGL works with a range of organisations that record London's species, from local 'friends of' groups and the majority of London boroughs to regional organisations such as London Natural History Society and London Wildlife Trust. See also this excellent website from the London Wildlife Trust, for projects with schools:

www.wildlondon.org.uk/Education/Servicesforschools/tabid/258/language/en-GB/Default.aspx

» *Set up your own biodiversity survey, relevant to your area, or (better still) work in partnership with a local biological records centre*

» *Work with local schools on indicators that you feel help to enhance local biodiversity and human understanding of its value, e.g. 'How many children have held a frog or worms?' or 'How many children can identify ten plants that are good for bees?'*

» *Ask local celebrities or councillors to become 'species champions' as a way of protecting and enhancing biodiversity in your area*

The plight of bees

Over the last few years honey bees have been mysteriously disappearing all over the planet; literally vanishing from their hives. The phenomenon has been dubbed Colony Collapse Disorder, but nobody knows why it's happening. Bees pollinate one in three bites of food on our plates, so this a crisis for us as much as for bees.

The problem had been seen as mainly concerning rural, and particularly farming, areas, which led to a focus on pesticides and monocrop farming, but as yet there is no conclusive evidence. As this book went to print a major study was about to start looking at whether a cocktail of chemicals from pesticides could be damaging the brains of British bees. By affecting the way bees' brains work, researchers say, the pesticides might be affecting the ability of bees to find food or communicate with others in their colonies.[11]

But it's not just bees in the countryside that are being affected. When beekeepers in North London opened their hives after the harsh winter of 2009/10 they found a large number of bees dead or disappeared, far more than could be explained by the cold. It seems clear that bees are in trouble everywhere, but nobody really knows why.[12]

When I stood down from Camden Council in 2010 I was working with the council's Biodiversity Officer on a three-point plan to raise awareness about the plight of bees inside the council, as follows:

- Screen the film about the plight of bees, *Vanishing of the Bees.*
- Install beehives on the roof of Camden Town Hall.

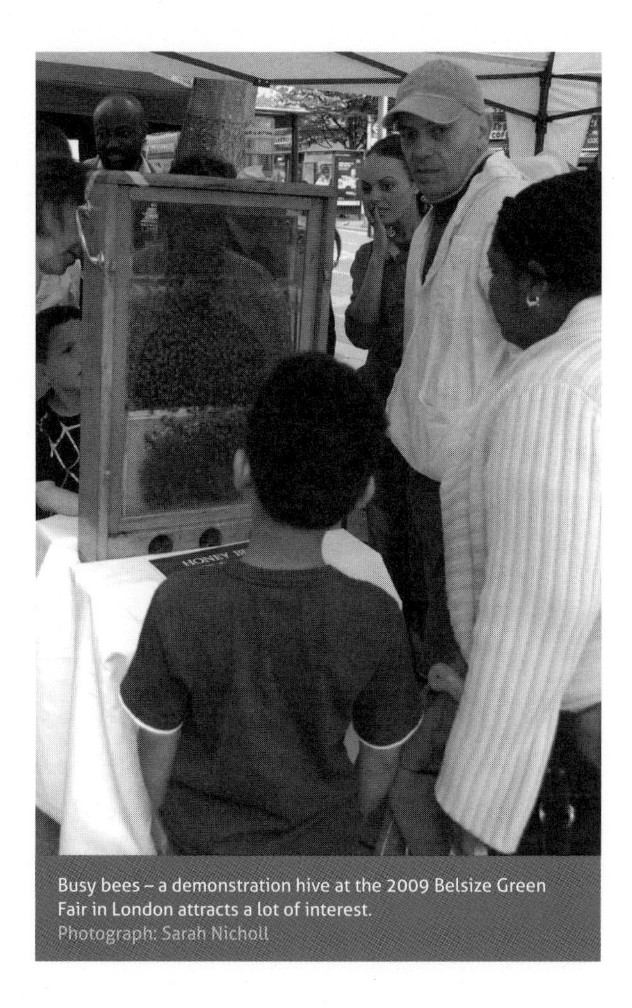

Busy bees – a demonstration hive at the 2009 Belsize Green Fair in London attracts a lot of interest.
Photograph: Sarah Nicholl

- Set up a course in conjunction with the North London Beekeepers Association so that council staff would be able to look after the bees on the roof.

In the end I ran out of time and it never happened, but it feels like an approach that could work, both in councils and in the community.

Chapter 5

ENERGY EFFICIENCY

Questions answered in this chapter:

- Where do you start on energy efficiency?
- What's the best way to create zero-carbon buildings?
- Is carbon rationing politically feasible and can it work locally?
- Who's done Energy Descent Action Plans and what role can councils play?

Energy efficiency is usually seen as boring because it's about insulation. Energy generation, by contrast, is exciting because it's about gadgets. But gadget-fixation can lead to classic mistakes, such as the installation of mini wind turbines on urban homes as David Cameron once tried to do – sometimes known as the 'eco-bling' approach to CO_2 reduction.

> "If [people] are putting costly photovoltaic cells, hot-water solar collectors and personal wind turbines on their becalmed, north-facing, turf-roofed, toxic timber-clad, non-airtight, poorly insulated houses, finished in high-emission materials – and they are sinking electricity-powered heat pumps and rainwater filtration systems down deep holes filled with recycled concrete, next to their reed-bed gardens with a focal-point bird table – then they could, just possibly, be spending money more wisely.
>
> The alternative to 'eco-bling', which is both cheaper and has much better green credentials, is 'eco-minimalism' – a good-housekeeping approach to ecological building design and specification, involving apparently non-glaringly obvious strategies such as insulation, draughtproofing and the use of healthy materials."
>
> Howard Liddell, Gaia Architects

Howard's main point is well made – we need to concentrate on reducing demand, not on eco-bling. That's partly about installing energy efficiency measures, which means knowing what exactly to do and how to pay for it; it's partly about education; it's partly about price; and it's probably, in the future, about carbon rationing.

There's not much that communities and local authorities can do on price (although if a council or a community group supplies energy services then it can have some influence – see Chapter 6, page 106), but there's a lot that can be done to help people to make their homes more energy-efficient, to raise awareness about why energy efficiency is important, and to prepare people for the possibility – nay, probability – of carbon rationing. There's also the Energy Descent Action Plan concept of the Transition movement, which is about a community planning the move to a post-oil future in their area.

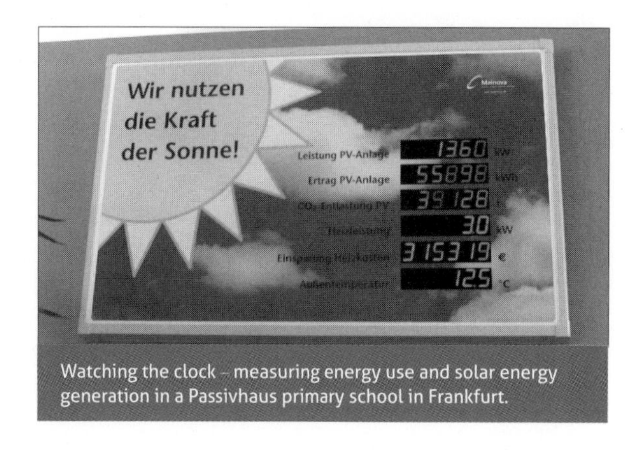

Watching the clock – measuring energy use and solar energy generation in a Passivhaus primary school in Frankfurt.

Where do you start on energy efficiency?

I'm constantly amazed by how the leaders on the climate change agenda in local government are doing what they think is right for their local community and the planet rather than what central government thinks is right. The following are some examples.

Revolving energy funds

Woking, a classic example of a district council punching above its weight and one of a number that feature regularly in this book, pioneered a system that was to revolutionise the energy efficiency debate. Back in the early 1990s Woking's then Finance Director, Ray Morgan (later Chief Executive), understood that investing in energy-saving measures would reduce his bills over time. But he also understood that, if he directed the energy bill savings into a fund, it would grow in time and provide more resources to reinvest. Hence the phrase 'revolving energy fund', which is sometimes also called 'invest to save'.

Many councils have now brought in revolving energy funds through the Carbon Trust's local government programme. Some councils have given additional incentives to departments to identify energy efficiency measures so that, for example, some of the savings are returned to that department rather than to the central fund. The Northern Ireland Office established a revolving energy fund for small and medium-sized enterprises. Derby set up a revolving fund for its schools. It's an extremely useful way to identify energy-saving measures and then fund them from future energy bill reductions.

Ask your council's energy department if it has installed voltage optimisation equipment, and if not, tell it about powerPerfector (see box)

powerPerfector at Oxfordshire and Camden Councils

Quite a few councils and other organisations have now installed a box of electronic tricks called powerPerfector, which optimises voltage, dealing with the discrepancy between the actual supply voltage received (207-253V) and the optimum voltage electrical equipment needs (220V).

Speedwell House is one of the main office buildings for Oxfordshire County Council. Like most office buildings, it has a broad mix of electrical loads, consisting mainly of lighting, HVAC (heating, ventilating and air-conditioning) and computers. powerPerfector reduced kWh consumption at Speedwell House by 17.9 per cent, equivalent to £4,303 and 36 tonnes of CO_2 saved per year.

» *Ask your council's energy or finance department if it has a revolving energy fund and, if not, urge it to set one up*

Draught Busting Workshops

In late 2009 Transition Belsize began offering free Draught Busting Workshops to teach residents how to reduce energy bills without having to install expensive double glazing and solid-wall insulation. The workshops were practical, hands-on events in local homes using good-quality window and door draught-exclusion products. I went to one of these and was (literally) blown away by how useful and friendly it was – 15 people passing on skills or learning new ones and working together happily as a community.

At the same time, Camden Council had won a bid to make Belsize and the surrounding area a Home Energy Efficiency Programme (HEEP) pilot for the whole of London. That meant up to 1,000 households each receiving ten energy and water efficiency measures and benefitting from an hour of energy efficiency advice. The aim of the funder, the London Development Agency, was to test the concept for the whole of London. Camden used it as an opportunity to identify residents who could be referred on to other schemes, such as boiler replacement or Pay As You Save (see page 92).

The Carbon Army in Camden

The British Trust for Conservation Volunteers (BTCV) is an environmental conservation volunteering charity. In November 2009 it launched the BTCV Carbon Army to coincide with the UN Climate Change Conference in Copenhagen, as a way of encouraging volunteers to take practical action on the environment. Carbon Army volunteers stacked up an amazing 6,000 days of practical climate change action during the time of the conference alone.

Camden's Carbon Army was launched in November 2009 when the kitchen garden at Lauderdale House, Waterlow Park, was recreated: planting herbs, fruiting shrubs and trees. Following the success of the launch, Camden Council commissioned the Carbon Army to deliver an ambitious project creating Community Orchards in nine of the most deprived housing estates and sheltered housing sites across the London Borough of Camden. Those involved in the project, which was launched in January 2010, spent two days a week through the winter planting the first ten orchards, which will produce fruit for the local population.

An average of 15 residents and volunteers per day from each estate and the surrounding area joined the sessions to prepare and plant the orchards. The emphasis of this project is on growing food, cutting down food miles and packaging, and promoting healthy eating, as well as developing communities and encouraging them to become more environmentally active.

See www2.btcv.org.uk/display/carbonarmy

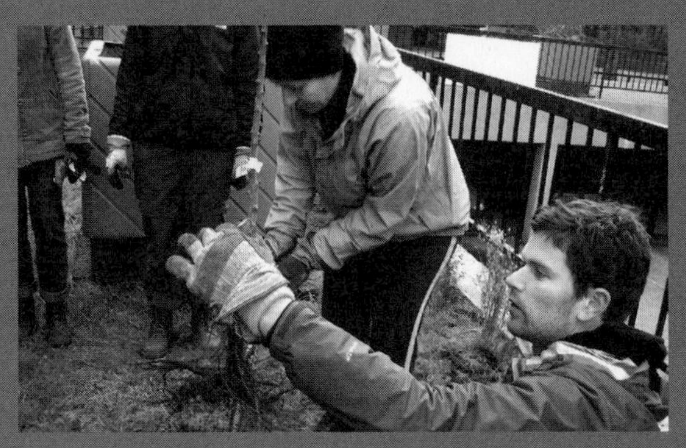

Chris Speirs and the BTCV's Camden Carbon Army planting orchards. Photograph: BTCV Carbon Army

The Draught Busting Workshops were a bottom-up attempt by residents to help themselves and enhance their community in the process. The HEEP pilot, on the other hand, was a top-down council initiative to install a range of measures, understand what other help could be offered and try to drive behaviour change. In an ideal world the council would buy the draught-busting materials, the council's HEEP advisers would refer householders to the local community draught-busting group, and the householder would learn to draught-bust his or her home in the company of others. And that is exactly what was starting to happen as this book went to press – Transition Belsize was doing draught-busting workshops paid for by the council. Hurrah!

» *Organise community-led draught-busting workshops in your area and ask the council's sustainability team if it can bulk-buy the materials for you or even donate them*

Community self-education and self-action groups

In 2010 Transition Town Totnes launched Transition Together (www.transitiontogether.org.uk), a project that aims to help small, social groups of friends, neighbours and colleagues take a number of effective, practical, money-saving and carbon-reducing steps. A workbook helps each person to build his or her own Practical Action Plan to improve household energy efficiency, minimise water use, reduce waste (and consumption), explore local transport options and promote the great-value, healthy food available locally. It also helps everyone to understand what's behind rising energy prices and climate change, and what this might mean for them, their family and their local community.

As one of the 2009 winners of the government's Low Carbon Communities Challenge, Transition Town

Community action – neighbours signing up to work together on a programme of environmental behaviour change in Totnes, Devon. Photograph: Fiona Ward

Totnes also launched Transition Streets, a scheme to help households install solar power at hugely discounted prices – thanks to a £625,000 grant from the Department of Energy and Climate Change (DECC), grants from South Hams District Council (SHDC) and low-interest loans from the Wessex Reinvestment Trust (WRT). Low-income households paid nothing up front – they received £3,500 from DECC, £1,000 from South Hams District Council and a loan from WRT. Higher-income households were eligible to receive only a £2,500 grant from DECC. To obtain a low-cost solar panel all households had to first go through Transition Together.

Global Action Plan's EcoTeams are similar to Transition Together in that they bring small groups of people together to work through a carbon-reduction programme. The key difference is that Transition Together addresses not only basic carbon-saving actions but also peak oil, community building and overall resilience-strengthening activities.

Global Action Plan had funding to be able to offer free EcoTeam leader training. It set a target of reaching 20,000 households over 2009-11 and ran leader training sessions across the UK. Anyone can sign up to join, or run, an EcoTeam, and the website allows groups to run on-line while also offering a range of downloadable resources. (See www.globalactionplan.org.uk/ecoteams.)

Carbon Conversations is an inspiring, practical, six-session course on low-carbon living based on the psychology of change. Led by trained volunteer facilitators, groups of six to eight members meet in homes, community centres, workplaces or other venues. The two-hour meetings are designed to engage people both emotionally and practically, helping them overcome the barriers often associated with making large carbon reductions.

Members explore the basic climate change problem, their responses to it, their ideas for a low-carbon future and the four key areas of the footprint – home energy, travel, food and other consumption. Most members make changes to their lives that amount to a reduction in annual CO_2 used of about a tonne, and develop plans to halve their carbon footprints over a longer period. See http://cambridgecarbonfootprint.org/action/carbon-conversations/national-carbon-conversations.

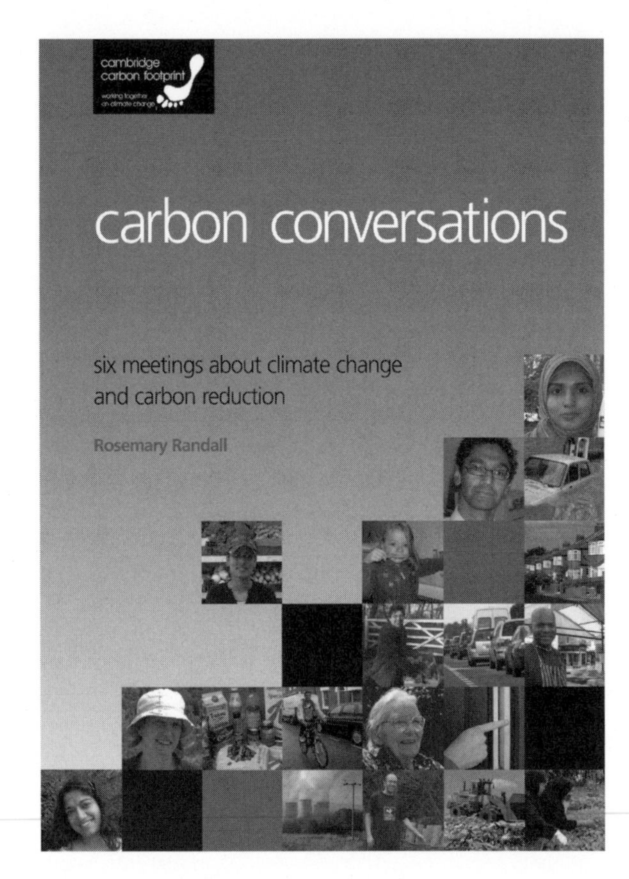

Roof and cavity-wall insulation

A national Warm Zones scheme was set up with government support in 2000 with the aim of developing new approaches to fuel poverty. 'Pathfinder' zones were established in 2001 in Hull, Newham, Northumberland, Sandwell and Stockton to trial different ways of tackling fuel poverty.

Much of the work was about installing measures – thermal insulation, draughtproofing and heating to improve comfort in the home. At the same time, advice on energy efficiency and benefits entitlement helped to reduce the amount spent on energy and maximise household income.

Warm Zones Limited (operated by National Energy Action, a leading fuel poverty charity and other partners) was set up with government help to manage the 'pathfinder' zones. The 'pathfinder' zone trials were completed in March 2004 and at that time full ownership of Warm Zones transferred to National Energy Action. There are now Warm Zones in London, the Midlands, Yorkshire, the north-east and Scotland. The largest Warm Zone programme is operated by Kirklees Council in the Huddersfield area of Yorkshire (see box).

Kirklees, Leicester and Sheffield councils have all committed to giving out free cavity-wall and roof insulation to all residents in both social and private housing. Some of the money comes from energy companies under the orders of central government (the Carbon Emissions Reduction Target scheme – CERT); some comes from the authority itself, because this is generally considered to be the cheapest way to reduce the carbon emissions of a local authority area.

The Kirklees Warm Zone scheme

Kirklees Warm Zone scheme is the largest local authority home insulation scheme in the UK. It offers free loft and cavity-wall insulation to every household in Kirklees. It aims to make every home in Kirklees warmer, more comfortable and less energy-intensive over a three-year period – in order to reduce energy bills and reduce carbon emissions.

The programme involves assessing all households, street by street within wards. The households are approached using contact with community and voluntary groups, intensive marketing and direct mail. Trained assessors carry out door-to-door visits to check insulation status. If additional insulation is required, a contractor installs mineral-fibre insulation in lofts and cavity walls, at no cost to the home owner.

The area-based approach has increased contractor productivity by 50 per cent, as geographically scattered installation approaches can incur significant travel time for surveyors and installers. Kirklees also uses the door-to-door visits to raise awareness of other council services, for example of benefits by vulnerable citizens, leading to significant increases in uptake.

In 2006 it was estimated that there were around 35,000-45,000 homes in fuel poverty in Kirklees. By June 2009 32,068 households had received at least one insulation measure. This is 51 per cent of all households surveyed by contractor; 85 per cent received loft insulation and 15 per cent received cavity-wall insulation. The insulation measures were estimated to save the average household £200 per year on their fuel bills.

The programme, which started in 2007, cost £20m, of which £11m came from Scottish Power as CERT funding and £9m as capital spending by Kirklees Council. Rather wonderfully, it's estimated that every £1 invested through the programme returns £4 to the local economy.[1]

See www.warmzones.co.uk/kirklees

Flying high – using abseilers to fill the cavity walls of 1960s council blocks. Photograph: Camden Council

Many urban councils have tower blocks that were built in the 1960s. They can have expensive cladding fitted externally to improve insulation, but they tend to have cavity walls so it would be cheaper to fill these if it were possible. It hasn't been possible until recently. Now Camden has pioneered a system of installing cavity-wall insulation in high-rise towers using abseiling techniques.

» *If your area has tower blocks or housing estates where residents have no control over heating, then suggest cavity-wall abseiling to your council's Housing Department*

Heat metering

Camden has been testing the installation of heat meters and heating controls on one of its housing estates in an attempt to give residents an incentive to cut their energy consumption as well as to introduce a fairer system of payment, i.e. pay for what you use. A lot of effort has gone into talking the project through with residents. Academics from Imperial College London and Oxford University have been conducting consumption/heating behavioural change research. The idea here is that it's social norms that govern behaviour – so, for example, if everyone on the estate is heating their homes to 20°C, then that becomes the social norm.

Solid-wall insulation

In a sense, cavity walls are easy, because they can be filled. The much bigger problem is what to do with those buildings that were built before the war and which have solid walls. They need solid-wall insulation, which means a thick layer of insulation on the inside or the outside of all exposed walls.

In 2008 Camden Council's Housing Department retrofitted a Victorian semi-detached five-bedroomed fourth-floor street property in a conservation area with a view to reducing its carbon emissions by 80 per cent. The initial target was actually 90 per cent, but the measures needed to get from 80 to 90 per cent were considered to be too expensive and difficult. Solid-wall internal insulation, high-quality replica Victorian double-glazed sash windows, solar water heating and localised heat exchangers were installed (as were photovoltaic panels, but only because they were given for free). Theoretically the energy bill for the house went down from about £2,000 a year to £200 a year.

I say theoretically because we now know that this is what might have been achieved if: a) the building fabric had performed as well as predicted; and b) somebody who cared about the planet had lived there. University College London researchers put monitors in the walls, which showed that the new occupants – three generations of a Bengali family from Camden's lengthy housing waiting list – set the thermostat at a

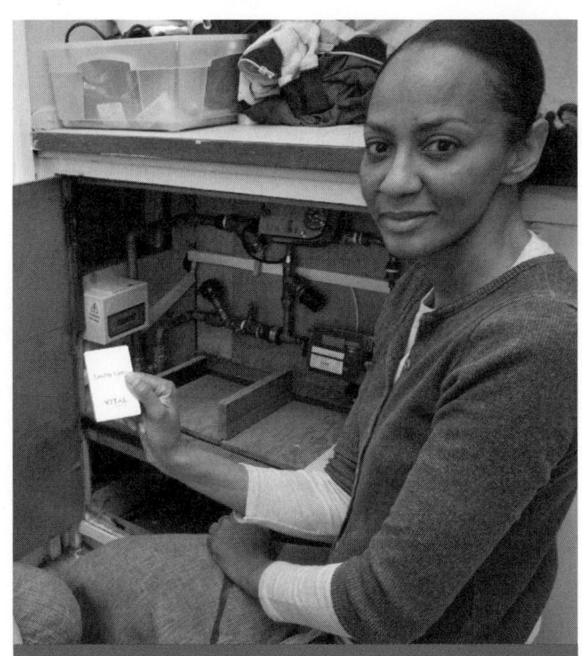

Feel the heat – a heat-metering pilot on a Camden housing estate. Photograph: Camden Council

staggering 25 degrees. In other words they took the energy efficiency gains in terms of higher temperatures – a real life example of the 'rebound effect' (see box).

Eventually the average winter thermostat temperature came down to 21 degrees after considerable education work was done with the occupants. That, plus the resolution of most building fabric and eco-bling issues meant the house performed better – achieving a 65 per cent reduction in CO_2 – but still not as well as predicted.

Overall, the Camden Eco House illustrated three key energy efficiency problems.

- There is a real need to train the building trade to do the work properly. They need to understand why draught-proofing is so important and why installing insulation to pass an airtightness test is not like slapping on plasterboard. Camden trained the project builders but the initial airtightness test still found a lot of leaks.

The rebound effect

The rebound effect was first proposed by William Stanley Jevons in 1865 – he argued that increasing the efficiency of steam turbines would increase, rather than decrease, the overall consumption of coal. As the cost of energy went down, he said, people would be more likely to use steam turbines more often. His prediction came true and the rebound effect was born.

An example of a modern rebound effect would be the driver who replaces a car with a fuel-efficient model, only to take advantage of its cheaper running costs to drive further and more often. Or a family that insulates their loft and puts the money saved on their heating bill towards an overseas holiday.

In 2007 the UK Energy Research Centre (UKERC) produced a report[2] that said: "If we do not make sufficient allowance for rebound effects, we will overestimate the contribution that energy efficiency can make to reducing carbon emissions. This is especially important given that the Climate Change Bill proposes legally binding commitments to meet carbon emissions reduction targets. We need to get the sums right."

The lead author of the report, Steve Sorrell, concluded that the rebound effect should be taken more seriously by governments when setting climate policy – in particular, making sure they focus on measures outside simple energy efficiency. "Our new understanding of the rebound effect reinforces the case for price-based measures, such as carbon taxes and emissions trading, to control emissions directly."

- Measured outcome in British low-energy homes that don't use a clear route map like the Passivhaus standard (see pages 94-96) rarely matches up with the predicted outcome. And very few post-retrofit houses are or will be measured.

- You have no control over post-retrofit behaviour by occupants so, without significant price rises by the energy market, there needs to be significant education of occupants on how and why to minimise their energy use.

The Camden Eco House is open a couple of times a year and is part of the Sustainable Energy Academy's Old Home SuperHome programme. The aim of Old Home SuperHome is to provide case studies on all significant retrofits (more than 60 per cent reduction in CO_2) in the UK in an attempt to inspire others to do the same. By the end of January 2010 there were 46 SuperHomes around the country. Most were private jobs but some were council or housing association properties. Wherever possible all the buildings in the SuperHome programme are also opened up at least once a year for public viewing. See www.sustainable-energyacademy.org.uk.

» *Lobby your council's Sustainability Team or Housing Department to create an exemplar retrofit building, or persuade someone in your community to retrofit his or her home and organise your own open days or join the SuperHome programme*

I think we probably missed a trick with the Camden Eco House in the sense that we didn't have the enabling mechanisms to get those who were inspired by the project to carry out improvements to their own houses. We did not tell them who could do the work, because frankly not many builders could, and we did not give them the financial mechanisms to help to pay for the work.

With cavity walls and roof insulation some councils have bitten the bullet and paid 100 per cent of the cost for everyone in social and private housing. But – and it is a big but – no local authority can afford to pay for solid-wall insulation and double-glazing out of their budgets. So what to do?

Pay As You Save

After two years of lobbying by myself and others the government launched a series of Pay As You Save pilots in 2010.[3] This is a variation on the revolving energy fund idea (see page 85) but aimed specifically at solid-walled homes in private ownership. Local authorities borrow long-term and then lend to householders wanting to install solid-wall insulation and double glazing. The loans are repaid either when the property is sold or out of energy bill savings. Stroud in Gloucestershire, Birmingham and the London Borough of Sutton (in cooperation with B&Q) are now testing the idea.

Installing internal wall insulation is most definitely a hassle for householders. But it's only the front and front-side facades of our older homes that need to be preserved for the sake of our heritage. The rear and rear-side can be fitted with external wall insulation, which would be much easier for residents and much better in energy efficiency terms if it covers up leaky brickwork. However, the plan would be to start with those who are really motivated to make environmental improvements to their homes. It seems fair to assume that they will be more willing to undergo the hassle factor than the general population. I would hope that by the time the roll-out goes mainstream there will either be such a financial incentive to do it that residents will accept the discomfort, or the process of installing the measures will have been improved dramatically, or we will have found a way to fit external wall insulation that meets the concerns of conservationists, as we have done with double-glazed replica sash windows.

In July 2010 I went along to Brent planning committee in north-west London to argue on behalf of two applications to retrofit pre-war houses with energy efficiency measures that would reduce the energy requirement of those buildings by more than 80 per cent. This is exactly the sort of thing we need to be doing. However, the buildings were in Brent's most best-policed Conservation Area. Conservationists, planners and councillors combined to frustrate the schemes despite the fact that it would have been impossible for passers-by to see the external wall insulation on the sides and rear of the houses.

The level of understanding among councillors about the need for energy efficiency was terrifyingly low. One councillor said the proposals would "make virtually no difference to the environment of Brent" [sic] and accused me of being "hellbent on destroying our heritage". It would have been funny if it hadn't been so depressing. And you couldn't have written the next part of the script. An application to demolish a listed building followed and that was passed unanimously! (See Chapter 8, page 131, for more on how planning committee was transformed in Camden.)

This tension between conservation and sustainability is going to be one of the biggest issues facing many parts of the UK over the next 20 to 30 years as we try to reduce the emissions of our housing stock by 80 per cent. We are at the beginning of a long and difficult journey.

» *If your area has a lot of pre-war, solid-walled homes, ask your council's Finance Department to follow up the 2010 Pay As You Save pilots and consider taking the same action*

» *Lobby your council's Planning Department to introduce the Uttlesford rule (see box)*

The Uttlesford example

A small district council in Essex called Uttlesford came up with a brilliant plan for encouraging energy efficiency through the planning system – it managed to establish the principle that a planning application can require energy efficiency improvements in another part of the building or another building altogether. Uttlesford said to homeowners wanting to build an extension 'you can only do that if you don't increase the overall carbon emissions of the building'. That's fantastic because it requires energy efficiency measures to be fitted to the main dwelling if the extension cannot be made zero-carbon.

So somehow Uttlesford established the principle that a planning application for one thing could require action in another area. In a sense we have this principle in planning with S106 agreements. The idea there is that if a planning application will create 'community bad', e.g. more traffic or more pollution or more pressure on school places, then the application should still go ahead if an agreement (called an S106 after the relevant legislation) can be found that alleviates those community bads. In the case of the small district council in Essex, they decided that increasing carbon emissions was a community bad so they found a solution.[4]

What's the best way to create zero-carbon buildings?

The 2005-2010 UK government set a goal of achieving zero-carbon new buildings by 2016. By mid-2010 it wasn't clear what zero-carbon meant – there was a consultation in early 2009 but it still hadn't reported back. But in a sense it's meaningless, because no house can be really zero-carbon when you take into account the construction and the living in it.

In 2009 I spent a lot of time and energy trying to persuade Camden to adopt the Passivhaus standard for new build and retrofit. I worked with architects, engineers, Passivhaus specialists and university researchers, who all committed a lot of time because they cared.

I persuaded two Camden councillors to go to a Passivhaus Schools conference in London in December 2009 and another to go on a Passivhaus trip to Austria.

The Passivhaus comfort & energy standard

Passivhaus removes the need for traditional active heating and cooling systems (boilers and air conditioning) by creating extremely well-insulated, draught-free buildings that optimise solar orientation and internal heat gains (from humans and electrical appliances). Passivhaus buildings include heat recovery ventilation to minimise winter and summer energy consumption, while avoiding overheating. The standard aims to reduce specific space-heating energy consumption to a maximum of 15 kilowatt hours for every square metre of floor area per annum ($15kWh/m^2/annum$) or a peak heat load of $10 W/m^2$). There must also be fewer than 0.6 air changes per hour (at a standardised pressure of 50 pascals), which is at least 15 times better than the average British Victorian house. (NB Any architect or engineer should be able to help you to make sense of those figures!)

There is a wealth of evidence from Germany and elsewhere to show that Passivhaus-certified buildings in Europe tend to deliver the predicted energy saving. This is by contrast with UK low-energy buildings, which have consistently failed to deliver against predicted consumption. UK regulations have focused on CO_2 as the compliance target, which rewards the installation of expensive renewable add-ons before reducing energy demand. The Code for Sustainable Homes, which is the main sustainability building standard in the UK, doesn't prioritise energy efficiency and is based on carbon reductions not energy requirement. So, under the Code, a biomass boiler is more acceptable than thick walls and triple glazing, despite the fact that the former can be easily replaced and the latter cannot.

At the end of 2009 the Building Research Establishment (BRE) released a 'Passivhaus primer'[5] that said it was impossible to reach the highest Level of the Code for Sustainable Homes – Level 6, which is sometimes called 'zero carbon' – without using Passivhaus design philosophy, which covers up to 85 per cent of the energy requirement of a building.

Wales leads the way – the devolved government has backed a plan to construct hundreds of Passivhaus buildings.
Photograph: bere:architects

I had already been to Frankfurt to learn how the city council had brought in the Passivhaus standard for all public buildings, even rented ones, and how that had encouraged the private sector to follow suit even before it was required to.

Camden's officers were concerned that "rigidly requiring the Passivhaus standard potentially prevented project teams from achieving the most cost effective solution to the first step of the energy hierarchy given specific building types and site conditions."

From a regulatory perspective, Camden's officers felt that there was "an insufficient evidence base to sustain a Passivhaus directive in the Local Development Framework under external review", which they felt was "reinforced by the fact that the government had yet to issue a policy statement on the zero carbon consultation". The real problem was that they were not prepared to go and look for the evidence base (or, perhaps more charitably, did not have the resources to do so). Dover District Council put together an evidence base to support its view that it needed to do more than central government was suggesting on renewables (see Chapter 8, page 128). That is seen as a beacon of good practice.

Frustrated by the unwillingness of Camden's officers to pick up the Passivhaus ball, I organised a free Passivhaus conference for London's planners, building control officers, architects and housing associations in February 2010. We were overwhelmed. All 160 places on the conference went within a few days. We could have filled Camden Town Hall several times over. It was clear that there was a real hunger out there for information about Passivhaus.

The keynote speaker, Liberal Democrat shadow Climate Change and Energy Secretary, Simon Hughes MP, vowed to become the first politician to live in a Passivhaus home. He also called on councils to introduce the Passivhaus standard into their planning rules.

I organised the conference because I wanted to dispel some of the myths around Passivhaus – like the idea that you can't open the windows, which was still doing the rounds at Camden Council in January 2010 – and because I want to encourage councils to put the Passivhaus standard into their planning rules. The UK building industry needs to acquire the skills, but that will come when planners, building control officers and architects start to feel more comfortable with Passivhaus. Camden Council's new Local Development Framework (LDF) encourages developers to use the standard, but it doesn't require it.

Three things happened as a result of the conference:

- I managed to persuade Camden's Planning Dept to send nearly 50 officers to look round two Passivhaus sites under construction in North London. On a later visit I took the Head of Planning and the Director of Environment.
- The largest political group in Camden, the Lib Dems, announced that they would be seeking to build Camden's new civic offices in King's Cross to the Passivhaus standard.
- Sustainability Officers in the Housing Department started working on a plan to Passivhaus a social housing block in Highgate.

» *Ask your Planning Department to consider incorporating the Passivhaus standard into the council's Local Development Framework*

Camden's planning rules and Passivhaus

"To ensure that developments firstly incorporate energy efficient design, we will require schemes to adopt appropriate energy efficiency principles as highlighted in paragraph 22.5 above. An example of energy efficiency principles is the Passivhaus standard. Passivhaus includes:

- very good levels of insulation with minimal thermal bridges;
- good utilisation of solar and internal heat gains;
- an excellent level of airtightness;
- good indoor air quality, provided by a whole house mechanical ventilation system with highly efficient heat recovery.

"The Council will strongly encourage schemes to meet the Passivhaus standard. Further details on energy efficient design and principles and Passivhaus are set out in our Camden Planning Guidance supplementary document."

Policy DP22, LB Camden Local Development Framework

Built to last – creating a Passivhaus wall out of wood in a factory in Austria. Photograph: bere:architects

If there's something you want to achieve for your whole area, like getting your council to introduce the Passivhaus standard for new and refurbished buildings, then educate yourself extremely well, start a campaign, find allies, get the press on your side, dispel myths, organise a conference in your Town Hall, keep improving your level of expertise and never give up because somebody will listen one day

Is carbon rationing politically feasible and can it work locally?

When David Miliband was Environment Secretary in 2006/7 he set up a committee to look at how personal carbon allowances might work. Sadly, his successor Hilary Benn scrapped it on the grounds that it was unworkable (by which he meant politically unfeasible), before it had a chance to evolve into a system that might work. One such might be Tradable Energy Quotas (TEQs), an energy-rationing system that was the brainchild of Dr David Fleming, one of the spiritual grandfathers of the Transition movement. TEQs (see www.teqs.net) are a way to plan the reduction in carbon by giving out fewer and fewer entitlements to energy over time and using the market to trade them.

We've already seen in this chapter that it's often not enough to do the technical work (e.g. installing energy efficiency measures) if there's no behaviour change or if the technical fix leads to rebound effects that can potentially worsen the situation. If councils or

David Fleming, inventor of Tradable Energy Quotas and wise elder of Transition Belsize, if not the whole of the Transition movement! Photograph: Sarah Nicholl

Voluntary carbon rationing in the community

A CRAG (Carbon Rationing Action Group) is an example of a voluntary form of carbon rationing. It's a group of people, usually in the same locality, who work together to reduce their carbon footprints.

CRAG members set themselves carbon allowances that are reduced year-on-year. Then, over an annual cycle, groups usually reward or penalise members that undershoot or exceed their allowance, putting excess funds into low-carbon projects or offsets.

The aims of the original CRAG, set up by Andy Ross in the West Midlands, were as follows.

1. To make people aware of their personal CO_2 footprints.
2. To find out if carbon rationing can help people make radical cuts in their personal CO_2 emissions.
3. To help people to argue for (or against!) the adoption of similar schemes at a national and/or international level.
4. To build solidarity between a growing community of carbon-conscious people.
5. To share practical lower-carbon-living knowledge and experience.

Each CRAG is self-starting and remains an autonomous entity that can adopt or set its own rules. For example, some groups (e.g. Leeds) have set individual, rather than common, allowances, and others prefer the less daunting title of 'Carbon Reduction Action Group'.

See www.carbonrationing.org.uk.

governments are to secure behaviour change, then they will need help from the community. Voluntary carbon rationing, as in Carbon Rationing Action Groups, or CRAGs (see box, right), is an example of how the community can pave the way for wider political action.

» *Join or set up a CRAG in your area and ask your council's Sustainability Team to publicise it*

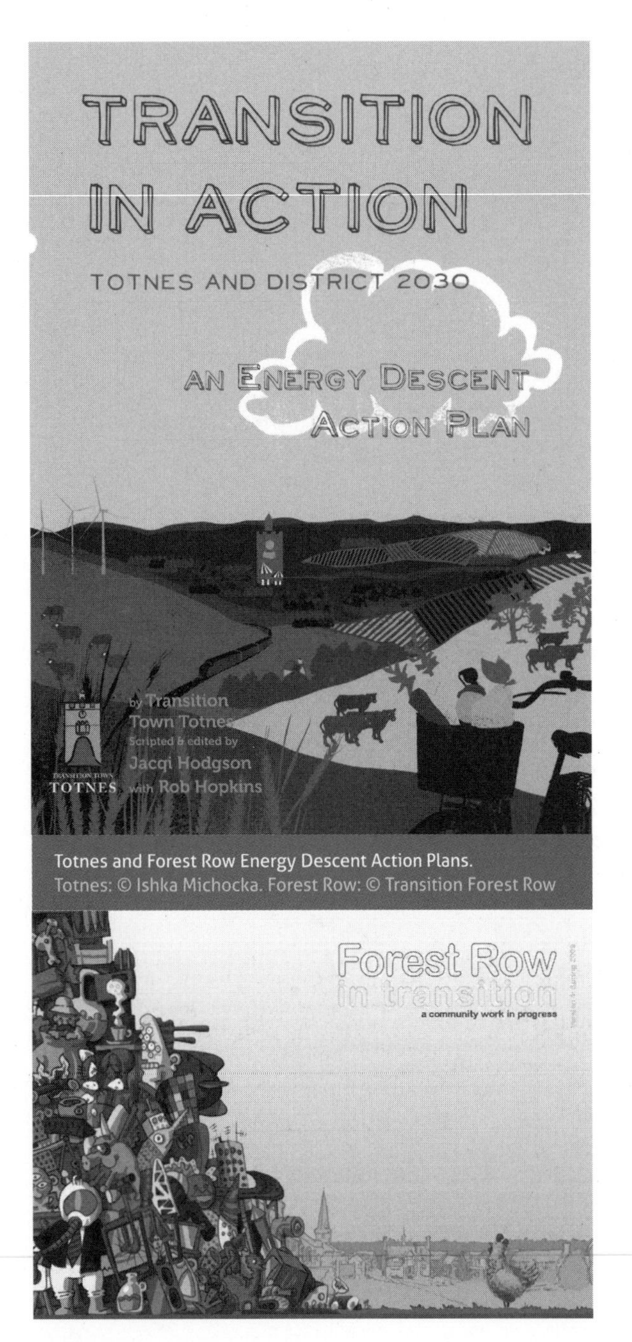

Totnes and Forest Row Energy Descent Action Plans.
Totnes: © Ishka Michocka. Forest Row: © Transition Forest Row

Who's done Energy Descent Action Plans and what role can councils play?

The last step of the original '12 Steps of Transition' described in *The Transition Handbook* is to create an Energy Descent Action Plan (EDAP) for your area. This is about getting members of the community to think through for themselves how they would like their low-carbon future to look. The key task is to design out the carbon, and the key assumption is that life with dramatically lower energy consumption is inevitable, but it's also supposed to be a positive visioning exercise and a plan for transforming all areas of community life for the better.

The first EDAP was done by students at Kinsale Further Education College in Ireland, which is where Transition founder Rob Hopkins was working as a permaculture teacher. The key lesson of permaculture is to use human ingenuity to replicate nature and design out energy and waste. That's what an EDAP tries to do. The Kinsale EDAP was an educational experiment. Since Kinsale, Totnes in Devon has published a serious EDAP,[6] while Transition Forest Row in Sussex has produced a mini-EDAP,[7] or maybe a halfway-house EDAP, which it has successfully used to raise awareness about Transition issues in its community.

> "Totnes is fortunate to have so many people who care about their community, who have strong views about how it should develop and what the priorities should be. Naturally there is a range of views but this can only be healthy for democracy. South Hams District Council welcomes all views but is particularly grateful to Transition Town Totnes for putting so much thought and effort into producing the Energy Descent Action Plan. It sees this as an important document in stimulating both debate and

consideration as to what all those involved in the life of the town can do to make Totnes an even better place to live, work or visit."

David Incoll, Chief Executive, South Hams District Council

Monteveglio and Bristol

In late 2009 Monteveglio Council in Italy announced that it was implementing an energy descent plan to turn Monteveglio into a 'Post Carbon' Town. It described the plan as "a strategic partnership with the Association Monteveglio Città di Transizione [Transition Initiative Monteveglio] with whom this administration shares a view of the future (the depletion of energy resources and the significance of a limit to economic development), methods (bottom-up community participation), objectives (to make our community more resilient, i.e. better prepared to face a low-energy future) and the optimistic approach (although the times are hard, changes to come will include great opportunities to improve the whole community's quality of life)."

Meanwhile, Bristol is now looking to build on the success of its Peak Oil Report, which was commissioned by its Local Strategic Partnership, and turn it into the beginnings of an EDAP for the entire city. If Monteveglio and Bristol are not perfect examples of communities and councils working together, then I don't know what is!

- *Create an EDAP for your area (as in Totnes and Forest Row)*
- *Talk to your council's Sustainability Team about creating a joint council-community EDAP for your area (as in Monteveglio)*
- *Try to persuade the Local Strategic Partnership to commission an EDAP for the area (as in Bristol)*
- *Think of a better name for Energy Descent Action Plan than EDAP!*

How many Transitioners does it take to draught-bust a front door?! Transition Belsize's Draught-Busting Group hard at work. Photograph: Sarah Nicholl

Chapter 6

ENERGY GENERATION

Questions answered in this chapter:

- What are decentralised energy and district heating systems?
- Does combined heat and power stack up?
- Which renewables make the most sense?
- How best can communities and councils supply energy services?
- Can you generate energy from food waste?
- Energy from waste – is incineration always a complete no-no?

What are decentralised energy and district heating systems?

Heat is produced as a by-product of electricity generation. In big power stations in the UK the heat is dumped into the air via large chimneys called cooling towers. Half the energy in a modern gas-fired power station is lost as heat. Older coal-fired power stations are even more wasteful, averaging just 37 per cent efficiency. Losses of a further 7-9 per cent occur as the electricity travels to users along the National Grid.[1]

Decentralised energy systems are where electricity is created close to or at the point of use so that the heat produced during power generation can be used locally. This is known as combined heat and power (CHP) or co-generation. Tri-generation, also known as combined cooling, heat and power (CCHP) includes absorption chilling (which works like a household fridge, by sucking warm air out of something or somewhere) in order to provide cooling in addition to heat. The act of supplying heat over an area-wide network is called district heating.

CHP systems are often powered using natural gas, which may seem counter-intuitive since it's a fossil fuel, but the point here is efficiency – they are meant to use much less fuel to deliver energy to end users.

CHP systems could be based on housing estates or they could be installed in neighbourhoods containing a number of energy-hungry institutions. The possibility then exists to link buildings to form district heating and cooling systems. The key is to mix different types of demand to balance loads. If a new neighbourhood is being built, then it may make sense to connect all the homes to the system, but it's unlikely to make sense to retrofit heat pipes to older dwellings.

District heating is not a new concept. The UK's first such system opened in 1950 in Pimlico in central London and made use of waste heat from Battersea power station. One of the UK's largest district heating schemes is EnviroEnergy in Nottingham, which was initially built by Boots the Chemist and is now used to heat 4,600 homes and a wide variety of business premises. The heat source is a waste-to-energy incinerator.[2]

In Denmark, district heating accounts for more than 60 per cent of space heating and water heating. In 2007, 80.5 per cent of this heat was produced on combined heat and power plants. Most major cities in Denmark have big district heating networks. The largest is in the Copenhagen area, which serves 275,000 households (almost 95 per cent of the area's population) through a 54km network of double district heating distribution pipes, providing a peak capacity of 663MW.[3]

Sharing the workload – the energy monitor for the Birmingham Energy Centre, the first of a ring of CHP schemes to be placed around the city.

There are an increasing number of CHP-based district heating schemes in the UK, including in Woking, Aberdeen, Southampton, Tower Hamlets and Manchester. Birmingham has launched one of the most ambitious schemes. The council is aiming to install a ring of CHP around the city, based on council properties or schools, so that most large businesses, council buildings and other organisations can connect to it. The first one linked up City Hall, the Concert Hall, the International Conference Centre and the Hyatt.

Does combined heat and power stack up?

However some are now questioning the value of CHP, partly on efficiency grounds and partly because of peak oil. In 2008 a Cambridge don called David MacKay published a book called *Sustainable Energy – Without the Hot Air*,[4] which has become a bit of a bible in the environmental sector. In it he challenged the conventional wisdom on CHP as follows:

> "When comparing different ways of using fuel, the wrong measure of 'efficiency' is used, namely one which weights electricity as having equal value to heat. The truth is, electricity is more valuable than heat. Second, it's widely assumed that the 'waste' heat in a traditional power station could be captured for a useful purpose *without impairing the power station's electricity production*. This is sadly not true, as the numbers will show. Delivering useful heat to a customer always reduces the electricity produced to some degree. The true net gains from combined heat and power are often much smaller than the hype would lead you to believe.

> A final impediment to rational discussion of combined heat and power is a myth that has grown up recently, that decentralising a technology somehow makes it greener. So whereas big centralised fossil fuel power stations are 'bad', flocks of local micro-power stations are imbued with goodness. But if decentralisation is actually a good idea then 'small is beautiful' should be evident in the numbers. Decentralisation should be able to stand on its own two feet. And what the numbers actually show is that centralised electricity generation has many benefits in both economic and energy terms. Only in large buildings is there any benefit to local generation, and usually that benefit is only about 10% or 20%."

This is controversial in the sense that CHP-powered district heating is now conventional wisdom for urban areas and is always more than just one building. I also think MacKay fails to include the community and resilience benefits of local energy generation, which are admittedly hard to quantify. But even if he is right on the efficiency numbers, there's a second elephant in the room – peak oil. Is it not a little short-sighted to create a CHP-powered heat grid fed by a resource

Using heat productively – a combined heat and power plant in Freiburg, Germany.

Micro-CHP in Islington

In November 2009 I went on a fascinating trip to the London Borough of Islington to see an eco-refurbishment of a small ground-floor flat in a Victorian conversion done by a private company called United House. It was a brave stab at creating a low-energy home, to be applauded because nobody asked them to do it. But it was also a typical British low-energy building – no route map, a fascination with gadgets and lots of trial and error.

They had installed a mechanical ventilation with heat recovery system that was far too big for the two-roomed flat and a micro-CHP boiler that gave off a constant noisy hum. These two boxes took up a good chunk of the tiny kitchen. The micro-CHP used natural gas to create electricity when there was demand for central heating or hot water. At £6,300 the micro-CHP unit was an expensive bit of kit that would take about 50 years to pay back – more if you add the cost of the condensing boiler that had been built in as a back-up!

that's set to become prohibitively expensive and then run out? So if your council does go down the CHP route, you'll probably need to help it to do some thinking about future fuel sources.

» *Ask your council's Energy Department if it has explored district heating systems, but ask also whether it is planning a sustainable future source of energy to run it*

Which renewables make the most sense?

We are only really at the beginning of the discussion about which renewables represent the future, and clearly every region will be different. (See Chapter 8, page 128, for how Dover District Council produced a renewables guide for its area as an evidence base to justify why it felt it should go further on sustainability than buildings regulations or government guidance required.) For a detailed discussion of where we are right now I recommend David MacKay's book. I do not plan to repeat his work here but I do want to mention a couple of technologies in passing.

Ground source heat pumps – as close as we get to a free lunch

Once demand has been cut and all possible energy efficiency work done, David MacKay comes out strongly in favour of ground source heat pumps, which are like back-to-front fridges in that they pull heat out of the ground in order to heat something rather than pulling the heat out of the inside, as a fridge does, to create cooling. Ground source heat pumps have valves that can break but their energy source – the sun-heated Earth – is sticking around – unlike fossil fuels. The latest Japanese heat pumps apparently generate up to five times as much energy as is needed to run the pump. In cities such as London there's rarely enough space to install heat pumps horizontally, and even vertically many of us will hit underground railway tunnels and the like, but for the rest of the country this seems like a perfectly sensible way to go.

Solar power

Germany has become the world's top photovoltaics (PV) installer, accounting for almost half of the global solar power market in 2007. The country has a Feed-in Tariff for renewable electricity, which requires utilities to pay customers a guaranteed rate for any solar power they feed into the grid. By the end of 2009 Germany had installed 8.3GW of PV, equivalent to about 1 per cent of Germany's electricity demand.[5] In 2008 the medieval town of Marburg in Hesse brought in a planning rule that required a minimum of 4m^2 of PV per roof, and at least 1m^2 of PV for every 20m^2 of roof, otherwise building owners would be fined.[6]

In the UK the solar power market leaders among councils are Kirklees and Braintree,[7] which were the first to offer incentives to residents to install solar

Solar City – Freiburg is the perceived market leader in Germany in terms of installing solar panels.

water panels. Kirklees also came up with an innovative scheme for lending residents up to £10,000 interest-free to install micro-renewables, called RE-Charge (see box). The loan was secured against the property and repayable on the sale of the property or on transfer of ownership. By November 2009 some 50 systems had been installed under the scheme.

I think Kirklees made two early-mover mistakes. It used its own capital to fund the interest-free loans without having any clear sense of when they would be paid back. It might have been better to borrow the money long-term and charge just enough to cover the costs of the interest payments. It also started with the wrong thing – renewables rather than energy efficiency. The priority should have been to insulate homes. That said, what it did was brave and paved the way for other councils to do the same and for the government's Pay As You Save pilots in 2010 (see page 92).

» *See if your council can be persuaded to incentivise tried-and-tested renewables, e.g. solar water heaters and ground source heat pumps, BUT only if residents have done the energy efficiency work*

After years of procrastination the UK government has now agreed to introduce a Feed-in Tariff (FiT – see Resources section), which will mean that those who produce electricity from renewables such as solar panels will get paid more for putting power into the National Grid than they have to pay to take it out. This is what has kickstarted the micro-renewables revolution across Europe, leaving Britain far behind.

» *Lobby your ward councillor to ask your council to offer roof space to community groups at a peppercorn rent so that they can install solar panels, generate electricity and benefit from the FiT*

Key lessons learned from the Kirklees RE-Charge scheme[8]

- Those applying for RE-Charge in year one had a reasonable level of knowledge about renewable energy and were keen to reduce their carbon footprint. However, when asked to indicate on the application form which technology they wished to apply for, many people ticked every choice. This is a good indication that there is a gap in knowledge about the application and suitability of technologies, even amongst people who are relatively energy-aware.
- There is a wealth of information available on the internet and in magazines and journals about energy and renewables, but it is difficult to filter this information and apply it to individual circumstances.
- The householder was required to obtain 'in-principle' consent from his or her mortgage provider to the council taking a charge over the property. A number of people found it difficult to obtain this. However, this is an administrative task that the householder is asked to perform; it can be time-consuming and would add to the cost of running the scheme if the managing agent or council undertook this.
- A policy was introduced to approve applications only from householders where existing charges were not personal. Personal charges over a property, for example resulting from relationship breakdown or personal loans, would incur legal costs disproportionate to the amount borrowed.
- Land Registry entries are often not up to date. This requires householders to provide additional evidence, for example where the mortgage has been paid off or a spouse has deceased. This has involved some 'tidying up' of the Land Registry by the council.
- Taking property charges can be complex and a 'one size fits all' approach is not appropriate.

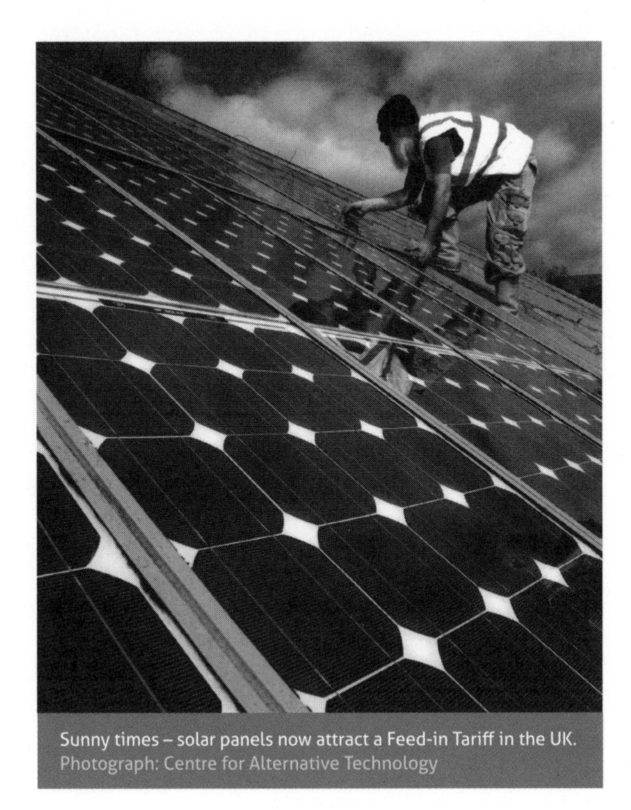

Sunny times – solar panels now attract a Feed-in Tariff in the UK.
Photograph: Centre for Alternative Technology

Moving from coal to biomass

In former coalmining areas such as Yorkshire and Nottinghamshire councils are seeking to move their buildings and schools from coal to biomass. Before it began its transition to biomass Barnsley Council was burning 6,500 tonnes of coal a year and generating 15,000 tonnes of CO_2 in the process. By installing wood heating in all new public buildings and refurbishments, Barnsley hoped to reduce its carbon emissions by 60 per cent. Sheffield City Council has converted the communal boilers in some of its tower blocks to biomass. In the process Barnsley and Sheffield have helped spawn a market for biomass fuel – wood chip – in what was once the heartland of the coalmining industry.[9] Nottinghamshire County Council did the same when it transformed all its schools from coal to wood-fired boilers.

NASA scientist James Hansen is on record as saying that coal is the biggest killer we face in the fight against global warming. He's right in so many more ways than he means. Coal is a killer of miners, both in pits and in later life; a killer of people through the inhalation of smog; a killer of plants and trees through acid rain; and now finally a killer of the human race because if we do burn all that coal that is left, then we will almost certainly trigger runaway climate change. Carbon capture and storage (CCS), the current techno-fix solution to the coal problem, remains unproven on an industrial scale.

» *Find out which organisations and businesses in your area are still burning coal and lobby hard to get boilers transformed to cleaner energy sources*

» *Collect waste wood from parks and open spaces to fuel biomass boilers (although not in urban areas unless the biomass boilers are equipped with filters or 'scrubbers')*

Wind power

Denmark generates a whopping 20 per cent of its electricity from wind. The UK, which has 40 per cent of Europe's wind blowing through it, obtained only 2 per cent of its power from wind in 2007, although the government is now seeking to ramp up the installation of offshore turbines.[10] Here's a big difference in approach between the two countries – virtually every wind farm in Denmark is a cooperative, which means the local community has a financial stake.[11] Bar some notable exceptions like the Energy4All wind farms (see below), no wind farms in Britain are cooperatives. In the UK large energy companies apply for planning permission to plonk mega wind farms in people's back yards. There's no connection with the local community. It should therefore come as no surprise when residents object.

How best can communities and councils supply energy services?

Energy4All was created in 2002 to facilitate the expansion of renewable energy cooperatives in the UK. It was a response to the phenomenal number of enquiries received by Baywind Co-operative from people looking to replicate the success of the UK's first community-owned wind farm, which was set up in Cumbria in 1996. Baywind generated enough green electricity to power 1,300 homes a year whilst paying an attractive return to its 1,350 members (averaging 7 per cent per annum), and supporting local initiatives.

» *If you want to set up a wind farm in your area, see Energy4All's Energysteps website – www.energysteps.coop – which provides information on the process of building a wind farm, then talk to the Energy Officer at your local council to see if they can help*

Blowing in the wind – Energy4All shows the way for community groups wishing to start wind farms. Photograph: Energy4All

Energy4All has done the heavy lifting for those wishing to set up a community renewables cooperative. Low Carbon West Oxford has set up an Industrial and Provident Society to rent roof space for installing solar panels and take advantage of the Feed-in Tariff. See box, right, for details of the West Oxford Community Renewables share offer.

Transition Maidenhead has set up a cooperative called Smart Energy to help people to navigate the minefield of renewable technologies and take advantage of the Feed-in Tariff. Smart Energy will find an approved installer in your area, negotiate group discounts and oversee the installation process. Depending on the area, it will also do the site survey and technical survey on behalf of the installer. Find out more at www.smartenergy.coop.

Many councils have set up Energy Services Companies or ESCos – usually joint ventures between public institutions and private companies – to deal with energy efficiency and energy generation infrastructure. ESCos started as vehicles for solving fuel poverty, but have developed into a standard method for using the know-how of the market to supply energy services whilst addressing the social and environmental concerns of the public sector. Woking, Birmingham, Southwark, Southampton and Aberdeen are among councils who have established ESCos.

ESCos can also be community-based and can encompass a wide range of activities beyond energy efficiency and energy generation. Four members of the Transition Town Lewes (TTL) Energy Group (Howard Johns, Liz Mandeville, Dirk Campbell and Chris Rowland) set up the Ouse Valley Energy Service Company (OVESCo) in 2007. They tendered for the contract to run the Lewes District Council grant scheme for domestic renewables and won (see box overleaf).

West Oxford Community Renewables share offer

"West Oxford Community Renewables (WOCR) was established for the specific purpose of building and owning community-owned renewable energy schemes in West Oxford and to thus generate funds for Low Carbon West Oxford. We will spend all the monies received from this share issue on renewables equipment and to construct and commission the renewable energy schemes, including solar panels on the big roofs in the industrial and retail estates, a micro-hydro at Oseny Weir and small wind turbines on Cumnor Hill.

The electricity generated will be sold to the building/land owners (with any excess sold to the grid) and the earnings will be donated to Low Carbon West Oxford to reinvest in community projects to make further cuts in CO_2 emissions.

Your investment in WOCR is in an Industrial and Provident Society for the benefit of the community, as withdrawable share capital. The full terms of the investment are set out in our offer document.

Your shares in WOCR should be considered to be a social investment, rather than one which will produce financial return. If none of the projects proceeds, your investment will be returned less any costs spent to date

This is a risk investment. Your shares will not increase in value and they may decrease in value. The Directors intend to pay a low rate of interest on shares, however it is not anticipated that this will be possible for the first five years at least. Any investment should be regarded as an investment for social and environmental benefit rather than financial benefit.

Your investment gives you a voting right which will allow you to participate in the operation of WOCR. The Society will operate on the 'one member, one vote' principle, whereby all investor members will have one vote regardless of the size of shareholding."

From the Low Carbon West Oxford website: www.lcwo.org.uk

CREATING A COMMUNITY RENEWABLES COMPANY IN LEWES

Chris Rowland, Director of OVESCo

Since 2007 Ouse Valley Energy Service Company (OVESCo) has been giving out grants in the Lewes area for renewables such as solar thermal and wood stoves on behalf of the local authority. The council awarded the contract to OVESCo because the company was set up by local people as a not-for-profit venture to benefit the community. This is an example of a council seeing the potential to tap into a Transition Initiative and make use of its members' skills and hard work to benefit all.

Since 2007 OVESCo has been awarded the contract to run the scheme from 2009 to 2010 as well as take on the role to advise the public as to where best to get grants for insulation. The work Transition Town Lewes (TTL), OVESCo and Lewes Council have done together has helped forge a good working relationship where each party tries to help the other achieve a common goal such as reduce fuel poverty, reduce local greenhouse gas emissions, and raise awareness about climate change and peak oil. It's important to understand that it takes the right people from Transition Initiatives and the council to make contact at the right time to make something like OVESCo work.

OVESCo has used its office to support TTL and work on projects to generate local sustainable power. OVESCo invested its hard-earned money with support from the council to undertake a feasibility study of the River Ouse to look at the potential for water power and is now looking at a couple of sites to generate power for around 50 homes.

Setting up a company has enabled TTL to get ideas off the ground and make projects happen, for example, two local Energy Fairs and an Open Eco House weekend. The council has also invited OVESCo directors and TTL to have a say in plans to adapt to climate change in the future, and in return the council can show that it is working with the community for a potentially sustainable future. The only failure is that with more funding and support TTL and OVESCo could achieve more and do so faster!

Chris Rowland, Director of OVESCo community renewables company in Lewes. Photograph: OVESCO

Can you generate energy from food waste?

Many councils now collect food waste, which comprises up to a third of householders' rubbish bins. Most is composted, but the government, in its 2007 waste strategy, recommended that councils investigate possibilities for generating electricity from organic waste via a process called anaerobic digestion (AD). This involves capturing the biogas from rotting organic waste to drive a gas engine, leading to the production of electricity. Alternatively, the gas can be cleaned and injected into the grid or used as vehicle fuel. Food waste, crop waste, sewage and animal slurries are commonly used on the continent to generate biogas, but up until very recently only the sewage plants in the UK have had any success in making a financial return.

Large-scale anaerobic digestion

In 2007 I visited Greenfinch, a DEFRA-funded food waste-to-electricity pilot in Ludlow in Shropshire (see www.biogengreenfinch.co.uk). At the time it was processing 75 tonnes of organic waste per week, most of it household food waste from Ludlow, Newtown and Somerset. A small proportion (2-3 tonnes per week) comprised grass cuttings from Ludlow.

Bags of food waste were macerated on arrival at the Greenfinch facility, and liquid waste from further down the plant added so the waste could be pumped. It took 100 days for the waste to go through the AD facility. Biogas generated from the process was turned into electricity in a gas engine, and the electricity generated ran the plant machinery. The surplus electricity goes to the National Grid. Liquid and solid digestate go to a nearby farmer. Greenfinch says it can be used as compost in gardens, allotments or parks.

The site footprint is not much different for 5,000, 10,000 or 20,000 tonnes per year and, if the collection vehicles are specified as electric vans, then the footprint is surprisingly small – around 40m^2. The plant has a single-skinned wall and there's a faint odour in the car park, but a double-door entry-and-exit system would remove that as an issue. This seems to suggest that AD can work in inner-city areas so long as the price of land is not prohibitive.

My conclusion is that all local authorities should be looking at AD because they all have food waste to deal with. So communities should be lobbying for action on AD. But there's also an attractive financial reason why councils should be clamouring for AD – many makers of AD equipment will offer a deal whereby if the council can provide the land, the AD equipment maker will pay the build cost and run the facility. They'll then charge a gate fee for organic waste and sell the electricity or vehicle fuel produced.

Using biogas to reduce smell in Exeter

Using biogas as a renewable energy source isn't a new phenomenon. Nearly 200 years ago street lamps in London and Exeter were powered using sewage gas as a way of reducing smells – burning it got rid of the noxious fumes. Nowadays it makes sense to burn methane rather than letting it enter the atmosphere, where it is 23 times worse as a greenhouse gas than CO_2, but 33 times worse if you include its indirect effect on tropospheric ozone and stratospheric water vapour.[12] So maybe we should go back to lighting our streets with sewage gas!

» *See if your Waste and Recycling Department has a plan for dealing with its food waste using AD and, if not, start an AD information campaign at your council and set up a visit to an AD facility for your ward councillors*

Small-scale anaerobic digestion

Companies such as Kompogas in Switzerland have been providing small-scale solutions (1,000 tonnes per year and upwards) for years, mostly for use in farms.[13] In India they've been generating biogas from crop waste and animal slurries for decades.

In 2009 Alara Wholefoods, a muesli company based on an industrial estate in Camden, began working on a plan to install a small-scale AD system. Alara had a track record on the sustainability agenda – it was the first company in Camden to go zero waste, and its founder, Alex Smith, has been slowly creating a forest garden on the industrial wasteland around his factory. The aim of the AD project was to collect food waste from companies on the industrial estate and a nearby housing estate, then to generate electricity from the biogas and use the compost for food growing in the area.

If this works, then we have a fascinating model for the future. Suddenly small-scale AD could be a solution on housing estates, industrial estates – anywhere really. In the countryside, where crop waste, food waste, animal slurries and sewage are potentially much easier to collect, there is huge potential for generating energy for farms and towns, as already happens in Germany.

Whether this is the future is of course an open question, but I think small-scale AD is likely to prove at least as interesting as huge energy-from-waste developments, which are justified using arguments of economies of scale and which create their own insatiable need for rubbish.

But for councils to be able to support small-scale solutions they need to have the intellectual capacity to

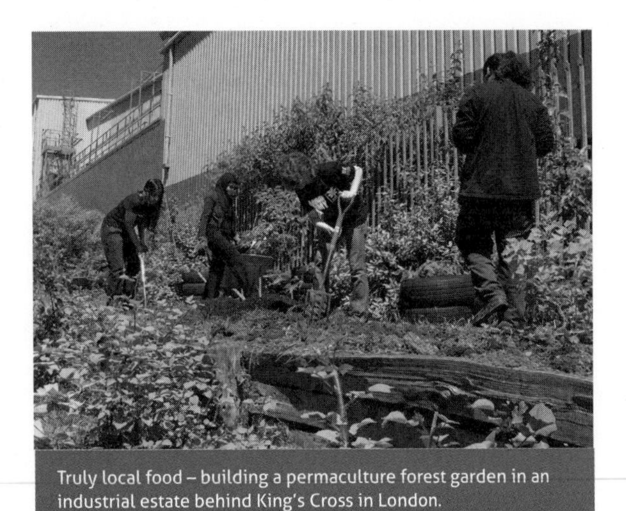

Truly local food – building a permaculture forest garden in an industrial estate behind King's Cross in London.
Photograph: Chris Speirs, BTCV Carbon Army

Biogas production in the German countryside

The small German town of Lunen was the first in the world to build and manage a biogas network. An anaerobic digester is fed with animal slurries and crop waste to produce gas, which is then used in combined heat and power plants across the town.[14]

Meanwhile, on the outskirts of a village called Jühnde in central Germany there's a biogas plant, a wood-chip-fired boiler and a biodiesel-fired peak load boiler – all connected to a new district heating network. The facility was installed after more than 70 per cent of the inhabitants agreed to change their energy supply from their own boilers to district heating from renewable resources.[15]

imagine a world where one size does not fit all, where community benefits can be measured alongside financial benefits, or where the politicians need to intervene – which almost certainly won't happen unless the community makes its views clear.

» *Watch what happens with the Alara AD project (see www.alara.co.uk) and ask your council's Waste and Recycling Department to join with you to investigate small-scale AD*

Producing vehicle fuel from food waste

In 2007 I visited a test facility on a landfill site near Guildford where a company called Gasrec was capturing gases created by rotting organic material and cleaning it up for use in vehicles. It claimed that using biogas or biomethane from food waste as vehicle fuel would lead to 80 per cent less carbon emissions than diesel and would mean zero noxious emissions. I was hooked and persuaded Camden to test the theory (see box, right).

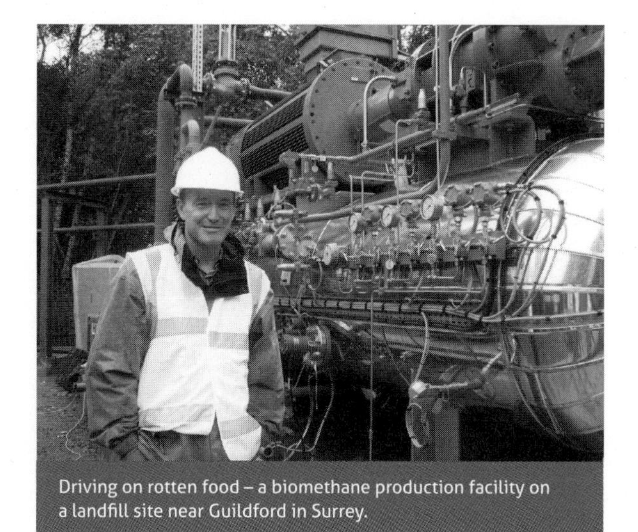

Driving on rotten food – a biomethane production facility on a landfill site near Guildford in Surrey.

The triple win that is producing vehicle fuel from food waste

In 2008 Camden became the first local authority in the UK to run part of the council's vehicle fleet on biomethane generated from food waste. There's now a biomethane refuelling facility in Camden's Transport Depot behind King's Cross. By coincidence, at exactly the same time the council's waste contractor Veolia was trialling biomethane in one of its rubbish vehicles being used in Camden and Westminster.

Biomethane is a triple win – it's made from food waste, it's low-carbon and it's good for air quality. In 2010 Camden was looking at a biomethane taxi trial. And the Mayor of Camden is being driven around in a biomethane-fuelled Volvo!

By the way, in case you're confused by the apparent conflict between methane being a powerful greenhouse gas and my excitement over burning biomethane made from food waste, I should point out that when methane is burned it produces carbon dioxide and water. Compared with other hydrocarbon fuels, burning methane produces less carbon dioxide for each unit of heat released.[16]

Sweden has already gone down the road of converting biogas to fuel for vehicles. With several thousand gas-powered vehicles and more than 50 biogas refuelling stations, Sweden is probably the most advanced country in Europe for the production of vehicle fuel from biogas.[17]

I personally doubt that this is quite the right model, although it's attracting a huge amount of investment in Europe. In my opinion there's never going to be enough biogas out there to keep the world' 800 million private cars on the road, at least not without converting food acreage to vehicle fuel crops, which would be a disaster. Furthermore, building a new filling station network is expensive. But there is likely to be enough biogas from organic waste to fuel municipal fleets, and council vehicles tend to stay in one area so they need only one refuelling station. In Lille in northern France about half of the buses run on biogas from local food waste and the aim is to have a 100 per cent gas-powered fleet by 2012.[18] Of course, if we all stop driving around in private cars and start car-sharing or joining car clubs, then there'll be a bit more biogas to go round.

Opportunities may arise in the long term for feeding biogas into our gas distribution network as a partial replacement for natural gas. I say 'partial' because there's going to be a lot of competition for biogas from crop waste, food waste, animal slurries and sewage. In Germany there's a Feed-in Tariff for biogas, which makes this form of energy production highly attractive. The Department of Energy and Climate Change produced a guidance note in December 2009 on injecting biogas into the gas grid, which said: "The Government intends to provide financial support for renewable heat under a new 'Renewable Heat Incentive (RHI)' . . . including how biomethane injection into the gas grid could be supported."[19] However, at the time of writing there were no further details available.

Setting an example – Camden's Mayor drives around in a biogas-fuelled car.

- *Encourage your council's Transport Dept to look at fuelling its vehicles with biomethane made from food waste*
- *Start thinking about how your community and council might take advantage of the planned biogas Feed-in Tariff*
- *Ask your council's recycling team to collect used cooking oil for turning into biodiesel (but please don't use the resulting fuel in urban areas! See box, right)*

Burning brightly – Amsterdam's AEB waste incinerator creates enough electricity to supply three-quarters of the households in the city. Photograph: AEB

Biodiesel from chip fat is bad news for air quality in urban areas

The London Borough of Richmond is one of a number of councils now running municipal vehicles on biodiesel made with used cooking oil. It seems self-evident that if catering establishments sent their used cooking oil to a company that turns it into biodiesel rather than pouring it into the sewers or sending it to landfill, then, on the face of it, there should be environmental benefits. However, I was concerned that the process of turning used cooking oil into biodiesel, which requires chemicals, might create more carbon emissions than the energy content of the end fuel. Camden Council therefore commissioned a lifecycle analysis of the used-cooking-oil-to-biodiesel conversion process, which showed that there were marginal carbon benefits to collection but significant air-quality problems. In an urban area therefore I wouldn't use biodiesel. To my mind the best place for it is in tractor engines on rural farms!

Energy from waste – is incineration always a complete no-no?

In November 2009 I went to see an autoclaving plant in Rotherham that quite simply steam-washes and blow-dries rubbish. The aim is to clean up waste to make it easier to separate out recyclables as well as create an organic residue (about 60 per cent) derived from the food waste, paper and card in the household rubbish. That residue can then either be anaerobically digested or used for soil remediation projects, and possibly one day used as a growing medium. Sterecycle, the makers of the world's first industrial-scale autoclaving plant, reckon that it will eventually be able to use 95 per cent of the incoming rubbish either as energy or as a resource material. There's also a lot of research going on into technologies for heating waste and extracting energy as a gas (gasification, pyrolysis, gas plasma), but they're still in their infancy.

The big question for those of us seeking the greenest possible solution for our waste then is this: do separation facilities or waste-heating technologies add benefit in terms of additional recycling, increased energy recovery and reduced carbon emissions, versus simply heating rubbish and extracting whatever energy we can? I look forward to someone doing the analysis because it's way beyond me!

Incineration has a bad name in the UK because we still assume that it means 1960s-style furnaces, which are inefficient and which spew dioxins and other pollution out into the environment. But we need to be careful not to compare pears with apples. Incinerators are not what they were. Amsterdam has a huge incinerator near the centre of the city that reckons to recover 99 per cent of the energy embodied in the waste it burns in the form of electricity (equivalent to 75 per cent of the power demand of Amsterdam's households), heat (which is used in district heating networks in the city) and other post-combustion products.

For example, at the Amsterdam incinerator metal isn't filtered out before combustion because the incinerator company AEB claims that doing it afterwards increases its quality, for example by eliminating food residues. And plastics are not recycled in Amsterdam because the analysis that was done suggested it was more cost-effective and better for the environment to burn them and create energy. The chimneys of the AEB incinerator are equipped with the latest air-quality scrubbers and output of airborne gases is monitored carefully by the city authorities.[20] So we need to be careful not to compare 1960s incineration with twenty-first-century incineration.

The bigger problem is the creation of monsters that need waste to feed them because economies-of-scale arguments lead inexorably to 'big is beautiful and cheaper'.In mid-2010 the North London Waste Authority was about to tender for what will be the second largest waste disposal contract in the country after Manchester's. The plan was to create some facilities for separating mixed recycling (materials recovery facilities, or MRFs) and some anaerobic digestion plants, and then to create something called solid recovered fuel (SRF) out of the residual waste. The SRF can then be incinerated somewhere – possibly at an industrial company or wherever there is demand. If the burning can be done cleanly, then there is even a possibility of doing it at the community level, somewhere like a housing estate, although I foresee a lot of resistance from residents to this sort of idea. There would also be the not-insignificant cost of complying with the Waste Incineration Directive, for which compliance costs can be up to £2m – not a viable proposition for a small-scale plant. However, the main issue is that by creating a waste-derived fuel rather than a huge incinerator North London is not locked into the need to keep generating rubbish.

» *Weigh up the pros and cons of incineration carefully if it's proposed for your community (and make sure a visit to Amsterdam is part of the consultation!), but don't dismiss it out of hand*

Chapter 7

FOOD

Questions answered in this chapter:

- Why do we need to reduce our consumption of meat?
- Why is fish even more of a headache than meat?
- How do we get more people growing food?
- How do we re-engage schoolchildren with the food chain?

We all have to eat. But very few people understand the extent to which cheap oil underpins our food system in the industrialised world and the impact the food system therefore has on climate change, as well as how vulnerable it is to peak oil.

Our food is dripping in cheap oil. Fertilisers, pesticides, tractors, combine harvesters, transport, packaging, refrigeration – without cheap oil our food supply is in trouble. Large cities are particularly vulnerable because they produce none of their own food, energy or water – everything comes from outside. If supply lines break down, then we will undoubtedly struggle.

Britain grows just 5 per cent of its fruit and 50 per cent of its vegetables. Numbers like that lead to starvation in times of crisis.

> "Our food security hasn't been threatened like this since World War Two – we are sleepwalking into a crisis."[1]
>
> Professor Tim Lang, City University, London

Reducing the carbon footprint of food will usually mean less food from far-off places, fewer out-of-season foods, no air-freighted produce, fewer processed foods, less packaging and less waste. It could be a lesson in permaculture – design out the energy needed and the waste produced. The bible on local food, *Local Food: How to make it happen in your community*,[2] was written in 2009 by Tamzin Pinkerton and Rob Hopkins, and I urge you all to read it.

Why do we need to reduce our consumption of meat?

Meat is the biggest food issue for those in the industrialised world, according to a UN study[3] – the livestock industry produces 18 per cent of the world's carbon emissions. That's partly because cows burp and fart methane, but also because of the fossil fuels that are used to grow grain to feed to cattle, to make processed feed-cake for cattle to eat, to pump water for cattle to drink, to refrigerate meat, to transport refrigerated meat, and then – craziest of all – to sell meat in supermarkets in open fridges and freezers and heat supermarkets to compensate for the chill. And the problem is set to get much worse as China and the rest of the developing world grow richer and consume more meat.

Vegetables, fruit, seeds, nuts, grains and pulses can provide all the protein, vitamins and nutrients that humans need. Indeed, for most of its existence the human race has primarily lived off this sort of diet. 'Meat and two veg' has only been a general cultural reality, as opposed to a story or aspiration, in the UK since the Second World War. Before that only the middle and upper classes ate meat regularly.[4]

It is only in the last 60 years, off the back of cheap food policies subsidised by the government, supported by the National Farmers Union and made possible by fertilisers and pesticides from chemical companies, that we have massively increased the quantity of meat and dairy we consume. And of course we now eat poor-quality meat, which often contains traces of antibiotics, growth promoters and pesticides, and we prepare it badly as well, often by frying it.

We compound the eating of poor-quality, badly cooked meat by eating it with huge quantities of highly processed foods made originally from plants such as soya but designed to maximise the shelf life of food products and the profits of food companies. We'd do well to cut down on all types of junk food but, as a start, eating less meat, but better-quality meat, will help both the planet and our health. In the context of localisation it should also help British livestock farmers because we will pay more for better-quality meat.

We need to change the prevalent cultural stories: the idea that food should be cheap, which became so ubiquitous in the post-war period; and that a proper meal is 'meat and two veg'. Poor-quality, cheap meat is almost always high-carbon and bad for our health. In short, our mantra should be 'less meat, better meat'.

However, less meat is a difficult sell. Or rather, it was when I started trying to do it in Camden in 2008. The proposal was blocked by the Camden Conservative group on the grounds that informed choice was a better policy than restricted choice. The *Evening Standard* and *The Guardian* both published stories suggesting I was calling for schools, care homes and council canteens to go vegetarian. The *Standard* even headlined its story: "Council's Green Advice to Staff: Go Vegetarian". My former employer, the Today Programme on BBC Radio Four, interviewed me in what might be called their 'look-what-crazy-councils-are-suggesting-now' slot. Apart from one enlightened farmer who wrote to me to say that he understood that 'less meat, better meat' meant more British meat, the UK livestock industry branded me a crazy bean-eater and wrote me venomous letters. Even more bizarrely perhaps, vegetarian groups around the world started contacting me to say thank you. Yet my view was, and is, not that everyone needs to stop eating meat, but that we need to eat less meat for environmental and health reasons, and that when we do eat meat it should be better-quality meat.

Thankfully, the tide began to turn in 2009 when one institution after another backed the 'less meat' concept. The NHS announced that it was reducing the amount of meat on hospital menus for environmental and health reasons. The first report of the UK's new Climate Change Committee, which advises the government on bringing down carbon emissions, said people needed to eat less carbon-intensive meat such as beef. The Chair of the Intergovernmental Panel on Climate Change (IPCC), Dr Rajendra Pachauri, was

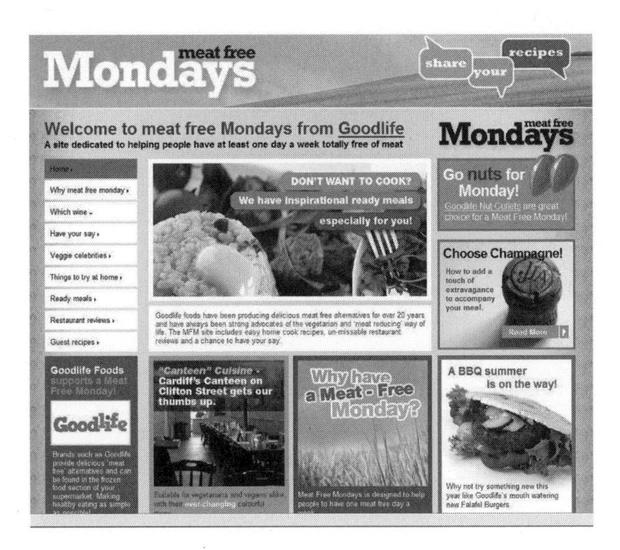

quoted as saying "meat production accounts for about 18 per cent of the world's total greenhouse emissions, so among options for mitigating climate change changing diets is something one should consider." Lord Nicholas Stern, author of the 2006 Stern Review on the economics of climate change, said "Meat is a wasteful use of water and creates a lot of greenhouse gases. It puts enormous pressure on the world's resources. A vegetarian diet is better."

Meanwhile the Belgian city of Ghent went largely meat-free on Thursdays and Sir Paul McCartney launched Meat Free Mondays, which some schools have now taken up.

» *Contact your ward councillors to see if they will help you persuade the council to introduce a 'less meat, better meat' policy*

» *Ask the schools in your area to introduce Meat Free Mondays*

Why is fish even more of a headache than meat?

Fish is an incredibly difficult subject for anyone trying to do the right thing. Overfishing is now widely recognised to be the single biggest threat to marine wildlife and habitats, and many fish stocks are known to be in severe decline. For easy reference when you're out shopping or at a restaurant, the Marine Conservation Society (MCS) has produced a Pocket Good Fish Guide. This wallet-sized list of the Fish to Eat and the Fish to Avoid is available free from MCS (see Resources section) on receipt of a self-addressed envelope.

Oily fish such as herring, mackerel or sardines are good for us, but most of us prefer white fish such as cod, haddock or monkfish, which are disappearing fast. Fish farms are mostly ecological nightmares – farmed salmon requires at least 3kg of wild fish for every kilo of salmon produced and usually contains traceable antibiotics and other cancer-inducing chemicals by the time it reaches your dinner plate. Industrial prawn farms in the Far East have decimated mangrove forests, which are some of the most productive ecosystems in the world. Big fish such as tuna contain dangerous levels of mercury and other industrial pollutants mankind has dumped in the sea. Line-caught fish – the most sustainable kind – are horrendously expensive. Sustainably fished stocks such as Alaskan cod have a high carbon cost because of refrigeration and transport.[5]

Marine Conservation Society fish to avoid and fish to eat list

Species to avoid: www.fishonline.org/advice/avoid

Species to eat: www.fishonline.org/advice/eat

Persuading a community to eat fish responsibly

Transition Belsize has a fledgling Sustainable Restaurants and Food businesses group. It was formed after a screening of the film *The End of the Line*, which tells the tale of how fish stocks have been decimated. The group has done excellent work with the Marine Conservation Society to persuade Budgens supermarket in Belsize Park to introduce a sustainable fish policy. All fresh and frozen fish is now colour-coded with a traffic light system – green dot is sustainably fished; red is not. Within a few months Budgens had stopped selling the red-dot fish altogether.

» *Organise a showing of 'The End of the Line', inviting your local councillors and someone from the council's Sustainability Team, then try to set up a group to educate local shops and restaurants about fish*

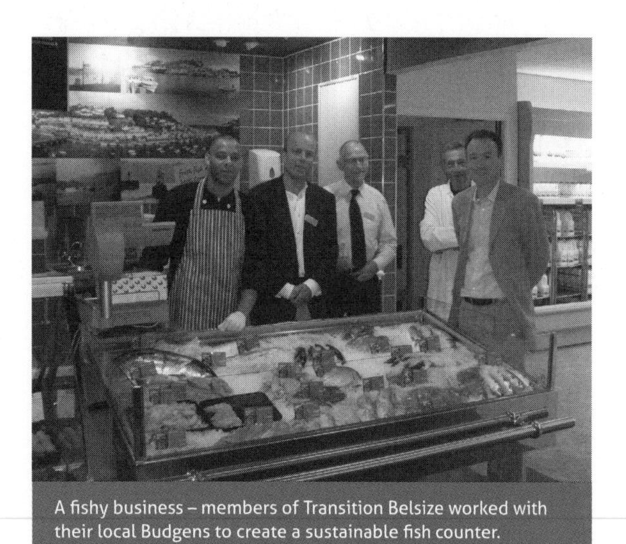

A fishy business – members of Transition Belsize worked with their local Budgens to create a sustainable fish counter.

» *Encourage as many people as possible to watch the following short lecture about the general destruction we have wrought on our marine environment: http://transitionculture.org/2010/07/21/jeremy-jackson-on-how-we-wrecked-the-oceans*

How do we get more people growing food?

Food growing is the key to many things. In a recession it's cheaper to grow your own. Food from your balcony or your garden or your housing estate hasn't travelled thousands of miles by plane wrapped in plastic and stifled by preserving gases. Gardening has been shown to be good for mental health. Growing food helps us all to stay in touch with where food comes from, and how important it is to respect nature.

Allotments

Many councils now actively encourage food growing. The 1908 Smallholding and Allotments Act requires local authorities to create allotments if approached by a group of six or more people in a borough. Inner London authorities are exempt from this because of the problem of space, but Islington is backing the trend by providing allotments and community growing spaces anyway (see box, page 122).

Under the 1908 Act those who rent allotments from local authorities are not allowed to sell their produce in local markets; they are allowed to grow food only for their families. However, a loophole appeared in the Allotments Act of 1922 that allows the sale of 'surplus produce'. In 2002 the government was reported to be looking at updating the 1908 Act with a view to increasing the local supply of food, but nothing came

Vertical food growing – it's amazing how much you can grow in a back yard. Photograph: Mark Ridsdill-Smith

of it. In 2009 Brighton submitted a request under the Sustainable Communities Act to lift the ban on allotment holders selling their surplus.

» *Get together with five or more wannabe food growers and ask your council's Allotments Officer for an allotment or, better still, identify a site which you think could be used for food growing and contact your council's Sustainability Team to see if it can help*

» *Follow the Todmorden example (see box, right) and either add food growing to your Community in Bloom competition or create a new, Incredible Edible Community competition instead*

The rapidly sprouting Incredible Edible Todmorden campaign

In 2009 the town of Todmorden on the Lancashire–Yorkshire border became the first in the country to supplement traditional floral displays, including hanging baskets and herbaceous borders, with slightly less colourful but more practical greens.

Vegetable beds, herb gardens and orchards have sprung up on sites as varied – and previously urban – as the railway station forecourt and an elderly people's home, under the aegis of the Incredible Edible Todmorden campaign.

Volunteers have replaced 'inedible' planting outside the town's health centre with apple and pear trees, made watercress beds in a local park and given free vegetable seeds to social housing tenants. Schools now try to use local produce and the long-term aim is complete self-reliance for food by 2018.

In January 2010 nearby Clitheroe and Rossendale were thinking about going the same way. The area has a strong tradition of allotments, recently bolstered by fruit- and nut-tree planting in public places by Transition Initiatives.

See www.incredible-edible-todmorden.co.uk

Incredible Edible Todmorden – turning public spaces into edible landscapes. Photograph: Estelle Brown

Food footprinting

In 2009 the first Transition Initiative, Totnes in Devon, published the results of several months' work on the question of whether the town could feed itself. Its final report makes for fascinating reading.[6]

If you assume that every back garden is producing food, and that the people of Totnes eat a lot less meat than they currently do, and drink a lot less alcohol, then the food footprint of Totnes – the area around the town needed to feed its 8,000 inhabitants – merges with those of neighbouring Torbay (130,000 residents) and Plymouth (241,000 residents). In other words Totnes can feed itself only if Torbay and Plymouth don't!

The most sobering food footprint is that of Greater London, which extends almost as far west as Bristol, and as far north as Birmingham. That's because Greater London has 7.5 million mouths to feed and produces virtually nothing itself. The *Can Totnes Feed Itself?* report points out that "Feeding the UK's cities will be a huge challenge, and raises many questions, including what degree of re-ruralisation will be required."

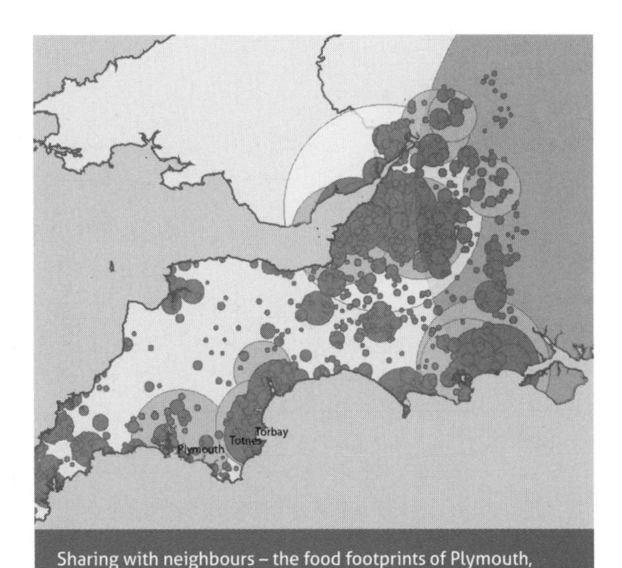

Sharing with neighbours – the food footprints of Plymouth, Totnes and Torbay merge, and London needs nearly a third of England to feed itself! Image: Geofutures

» *Create a food footprint of your area and then use it to lobby your council for more growing space*

Urban food growing

Cuba should serve as an inspiration to us all on urban food growing. When the Soviet Union collapsed in 1991 the Russians stopped supplying oil to Cuba and it became the first industrialised society to go through life after oil. Havana currently produces more than 50 per cent of the fruit and vegetables its residents eat. Could London or Manchester or Birmingham do the same? We won't know until we try, but Middlesbrough and Islington have shown the way (see boxes on pages 121 and 122).

» *Lobby your council's Environment Department or Sustainability Team to start a serious food growing programme*

Community land banks and land-share projects

Community land banks are one interesting response to the massive increase in interest in food growing but lack of allotment sites.

> "The concept is simple. The Community Land Bank would negotiate for land, hold it and then release it for rent to user groups under legally enforceable contracts, attracting charitable funding as appropriate, and facilitate transfers

From Robocop to radish crop

In 2007 Middlesbrough was the venue for a fascinating experiment in urban food growing. During the summer and autumn thousands of residents participated in a project to increase local food production and make locally grown food available to local people. The project was led by Middlesbrough Council and involved schools, allotment associations, mental health units, residents' groups and voluntary organisations. Residents and community groups were asked to identify unused land that could be used for food growing. If it belonged to someone other than the council, then officers negotiated with the landowner on behalf of local residents. The council also offered other enabling services such as advice workshops, free seeds and free tools for those on low incomes.

More than 1,000 people began growing fruit and vegetables in containers at 152 locations across the town. The experiment culminated in a town meal for 6,000 which cost just £80 – the price of the pasta!

Middlesbrough's food-growing experiment did two things. First, and perhaps most importantly for the town itself, it dispelled the image of Middlesbrough as a centre of antisocial behaviour, where 'Robocop' Ray Mallon was elected mayor following his zero-tolerance campaign as Detective Superintendent. There were only two incidents of antisocial behaviour against the food-growing sites in the first year – possibly because so many schoolchildren and their families were involved.

The second thing Middlesbrough did was replace Cuba as the pin-up of urban food growing. This is not to denigrate what the Cubans managed to do without oil, which was truly amazing, but it's sometimes hard to persuade people in an advanced capitalist society like ours that we should follow Communist Cuba, even when they're right! Middlesbrough gave those of us who were championing urban food growing a huge helping hand.[7]

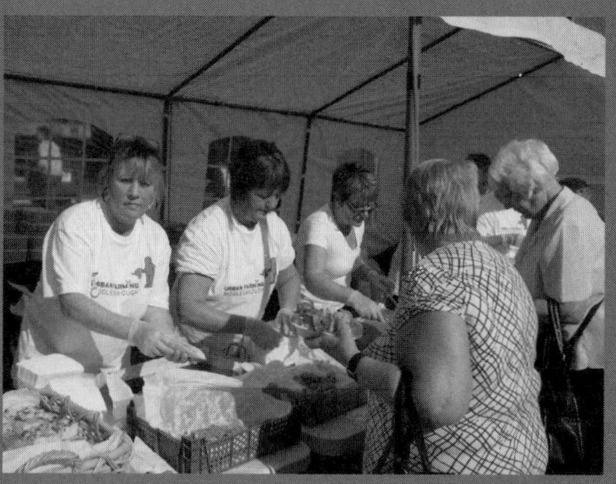

Feeding the 6,000 – Middlesbrough leads the way on urban food growing and translating that into a town meal.
Photograph: Middlesbrough Town Council

of tenants (community gardening groups) across a portfolio of land holdings. The Bank would also arrange insurance and ensure legal and technical compliance. In effect, it would be a safe pair of hands in which both landowners and users could trust.

A Community Land Bank would bring more land into cultivation and give local people the opportunity to get their hands dirty and grow their own produce. It would also give community groups the help and support they need not only to get their food-growing site off the ground but also to make it sustainable in the future and better able to avoid the pitfalls that so many fledgling community groups face."

Jeremy Iles, The Federation of City Farms and Community Gardens

Edible Islington

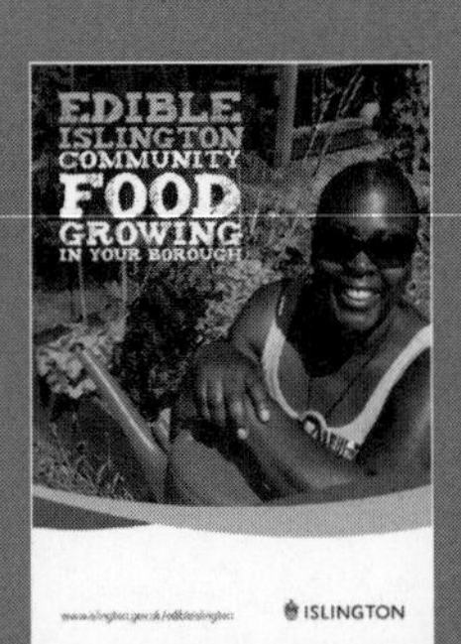

Edible Islington is a £1m capital fund founded in 2009 by Islington Council to establish and support the development of at least 60 new community food-growing spaces. Projects have to be delivered by a constituted group, not an individual, and must also incorporate water harvesting.

The fund employs three officers and is overseen by a project board comprising key borough partners and a steering group made up of the officers who have overall responsibility for the different programme components.

Grants are either large (£3k to £15k) or small (£300 to £3k). The large grants are for exemplar projects where community groups can prove that there is real need for a larger-scale food-growing project. The Edible Islington officers directly support community groups with large grant application and delivery. All large grants go to a steering group for a decision.

As part of the scheme Islington is planning to establish three new allotment sites, the first new allotments in the borough for a long time (and in an inner London borough that is under no obligation to provide allotment sites!). One of them involves digging up a car park, which feels like some sort of wheel coming full circle. The council are also working with the London Orchard Project to establish five new community orchards as part of the Edible Islington programme.

The fund is working with Garden Organic to establish a network of 40 Master Gardeners, who will receive accredited training and will be committed to in turn train at least ten additional individuals or community groups. In 2009 it funded a borough horticultural show run by the Islington Organic Gardeners Network and will also fund the show in autumn 2010, which will serve as a celebration for the achievements of the groups supported through the programme.

See www.islington.gov.uk/Environment/sustainability/sus_food/edible.asp

Land-share schemes seek to match up those wanting land with those who have land available. Garden share is a rapidly expanding small-scale version of the land-share concept. See the Resources section for examples of schemes and projects.

Transition Town Totnes has set up a scheme called Totnes Healthy Futures, which will give GPs the chance to prescribe community gardening to patients for health reasons. At the moment it's grant-funded but the aim is to turn it into a self-standing social enterprise. The founder of the Transition movement, Rob Hopkins, has also talked to police, both in Totnes and in the Forest of Dean, about whether they might promote garden share as a part of their community policing effort.

Fruit- and nut-tree planting

In 2007 I tried to persuade Camden Council officers to allow fruit and nut trees to be planted whenever a tree had to be replaced. "Not possible," was the reply. "It's a health and safety issue. Fruit might fall in the streets, people might slip, children might climb trees and fall out, the council would be sued, youths might use the fruit as missiles and anyway the trees would be destroyed by vandals."

Contrast that with southern Spain, where there are thousands of orange trees in public spaces and anyone can take the fruit. Chestnut and walnut trees are commonplace in French streets. Meanwhile Totnes in Devon has set itself the exciting ambition of becoming the 'Nut Capital of Britain' (see box, right) with the full support of the town's mayor.

» *Ask your council's Tree Officers to give out free fruit and nut trees or to require them to be planted when residents or developers want to cut 'troublesome' trees down*

Fruit for the future – planting fruit trees in Somerset.
Photograph: Andrew Norris

Organise your own fruit- and nut-planting scheme and set up a system for harvesting the crop and making good use of it

Farmers' markets

The railway lines that leave Camden's mainline stations – King's Cross, St Pancras and Euston – all pass through farming areas. The same is true of the trains running through Transition Brixton's patch in South London.

It would be great to forge links with farms along those lines with a view to supplying farmers' markets at the stations in Brixton and Camden because we're never going to be able to produce our own food. There's a farmer's market in my ward, which is great, but the farmers come in diesel-powered transit vans causing pollution and clogging up streets.

Councils can help by supplying space and giving planning permission and by talking to railway companies. Your group can help by talking to farmers and by getting the community on board to explain why a farmers' market is better than a supermarket.

'Totnes needs more nuts'

On 1 March 2007 the Mayor of Totnes, Pruw Boswell, led Totnes in its first steps towards its ambition to become the 'Nut Capital of Britain'. Transition Town Totnes and the Dartington-based Agroforestry Research Trust launched a new initiative to encourage widespread planting of nut trees in the area. Enthusiastically launching the initiative, the Mayor planted two walnut and five almond trees on Vire Island in central Totnes.

Rob Hopkins explained why nuts would always be welcome in Totnes: "The Transition Town Totnes project is already mobilising a movement to help the community to move away from dependency on oil for food, energy, housing and all aspects of our lives. If we can produce and enjoy much of our food locally, we can reduce the need for oil to transport our food. Not only will this lessen our impact on climate change, but having strong local food systems will protect our food security as oil prices rise. Nuts can provide an important local source of protein – especially for vegetarians – while simultaneously locking up carbon and providing very high-quality timber."

After 15 years of research, the Agroforestry Research Trust now has varieties of walnut and sweet chestnut that will crop reliably and heavily in our climate every year. It may be just a shell of an idea, but Transition Town Totnes hopes to see the community taking advantage of this pioneering work, and going nuts for the plan.

"Totnes needs more nuts," said the Mayor. "I am only too happy to do my bit." Adding "it takes one to know one!"[8]

See here for a video about nut-tree planting in Totnes: http://transitionculture.org/2010/03/01/a-short-film-about-the-totnes-nut-tree-project.

Work with your council's Markets Officer and regional farmers to set up farmers' markets and explore the options for bringing in produce by rail with the council's Economic Generation Team

How do we re-engage schoolchildren with the food chain?

Celebrity chef Jamie Oliver brought the parlous state of UK school food to the public's attention. Jamie's media profile forced the government to act where food campaigners had struggled to make headway. National Nutritional Standards, abolished under Margaret Thatcher, were reintroduced in 2007. School caterers were obliged to offer more healthy options. School kitchens, which had become microwaving facilities following compulsory competitive tendering in the 1980s, were rebuilt to permit cooking from raw ingredients. Kitchen staff were trained (or retrained) to cook.

Yet in the years After Jamie there was a reduction in children eating school meals, an increase in packed lunches and eating off-site and, for those continuing to eat school meals, a refusal among those of secondary school age to choose healthy options. And the figures for childhood obesity are alarming. By 2007 nearly one fifth (18 per cent) of three-year-olds in the UK were overweight and 5 per cent were obese.[9] Among 2- to 10-year-olds, 14 per cent of boys and 13 per cent of girls were found to be obese in 2008. For 11- to 15-year-olds, the figures were 21 per cent for boys and 18 per cent for girls.[10] Obesity-related diseases such as type two diabetes, which were previously unheard of in children, are starting to cause concern. Although the rise in childhood obesity rates has been less dramatic in recent years, the current prediction is that 30 per cent of boys and 27 per cent of girls will be clinically obese by 2020.[11] Bluntly, we are seriously compromising our kids' health by letting them choose junk.

Where did we go wrong? Well, here are a few reasons:

- Structural obstacles in schools. These include reduced budgets, minimal facilities, untrained cooks, uninspiring dining facilities, junk food outlets near schools, poorly written contracts, Private Finance Initiative issues and poor nutritional standards.

- Big business selling to children and creating confusion between wants and needs.

- The prevailing food culture in the UK, which is that food should be cheap, a meal is meat and two veg, and food comes from supermarkets.

Tackling the Takeaways – a Waltham Forest case study

As part of the consultation work around Waltham Forest's Sustainable Community Strategy the Council engaged with over 2,500 local residents. The feedback showed significant dissatisfaction with both the large number of local hot-food takeaways and their impact on the vibrancy and quality of town centres.

In March 2009, the council adopted a Hot Food Takeaways Supplementary Planning Document (SPD).[12] This statutory policy focuses on preventing any new Hot Food Takeaway Shops operating in inappropriate locations (e.g. near to schools, youth facilities or parks or in clusters around existing hot-food takeaway shops). The policy also outlines that as part of any planning application for new food takeaways the following must be considered: litter, waste management, odours and cooking smells, protection of residential amenity and hours of operation.

The Tackling the Takeaways strategy seeks to reduce the likelihood of childhood obesity through the implementation of policies limiting the development of hot-food takeaways within close proximity to schools and youth facilities. It's also one of those exquisite moments when a council does something you always think would be wonderful to do but impossible to realise!

- The lack of connection between children and the food chain.

» *If your area has a junk food takeaway problem, ask your council's Planning Department to prevent the creation of new outlets using Waltham Forest's SPD (see box, opposite)*

Here are ten things you could push for in your school:

1. Food growing and livestock rearing.

2. Treatment of food waste via wormeries or compost bins.

3. Visits to farms and food-processing facilities.

4. Cooking lessons and the chance to prepare the school lunch.

Carrots ahoy – food growing at Holy Trinity Primary School in Camden. Photograph: Camden Council

5. Redesign of dining rooms by pupils to create an environment in which they feel more comfortable.

6. Pupils kept in at lunchtime for sit-down meals, with staggered sittings if necessary, or replacement of cash with swipe cards to prevent pupils spending lunch money in junk-food outlets near schools.

7. Use of points and prizes to influence healthy-eating choices or (better) the removal of unhealthy options from school menus entirely and promotion of healthy options.

8. Introduction of stricter guidelines as to what is acceptable in packed lunches or even a ban on packed lunches altogether.

9. Meat-free Mondays.

10. Increased provision of education about nutrition for children and parents and for the organisers of breakfast clubs.

Of these, the key is really food growing. The rest will sprout (if you'll pardon the pun!) from that. The more we're able to connect our children with the fresh food chain the more they're likely to grow up with a taste for fresh food, rather than processed products, and the more they're likely to want to grow food on window-sills, balconies, in front yards and back gardens. And that's good for the planet and good for their health.

Chapter 8

PLANNING

Questions answered in this chapter:

- What do you *need* to know about planning law?
- How do you organise a campaign against (or in favour of) a planning application?
- Who can help you with planning issues?
- What should you push for in your Local Development Framework?
- Is carbon offsetting ever acceptable?
- What does positive collaboration between a council and the community on planning look like?

Planning is one of two areas of the council where community groups need some expert help. The other is procurement (see Chapter 9). Together these two services can be a huge lever for change. But they remain behind closed doors for many people – even for elected councillors.

What do you *need* to know about planning law?

It's perfectly possible to teach yourself the ins and outs of planning law but it takes a long time. Development Control Committee (the formal name for planning committee) meetings take place roughly every three weeks and last for up to three hours. Add to that the time you need to spend trying to understand the relevant parts of planning law and of particular developments, as well as the time needed to prepare deputations. It is a major commitment but definitely worth it if you can spare the time.

> "In the early stages of Transition Whitstable, one person on our steering group in particular made a lot of efforts to liaise with local council and councillors. Some members of the elected local council responded very positively and so did the local press. However, Canterbury City Council has proved a far more difficult nut to crack. We had instances a few years back where planning permission for private micro-generation, particularly solar, was being refused for what many regarded as paper-thin reasons. Planning permission for some eco-dwellings, notably one earth-sheltered dwelling, was also turned down. All this happened while major 'developments', some of these on greenfield

sites, continued apace, both residential and commercial.

On a brighter note, since Transition Canterbury has joined us as one of our sister Transition Initiatives, we now feel we have a lot more clout. Some of the Canterbury people are very active and articulate and have experience in liaison with official bodies such as the City Council. There have also been some other hopeful signs recently: a planning application to extend a holiday caravan site along the seafront was turned down due to public pressure. There has also been some airing of an idea to create a nature reserve on some very low-lying marginal agricultural marshland to the western side of the town, which would also be a natural soak-up zone in the event of surge tides up the Thames Estuary as well as being a landfall on key bird migration routes and a place for walkers and nature-lovers to enjoy."

Mari Shackell, Transition Whitstable, Kent

I was on Camden's planning committee throughout my four-year stint as a councillor, but by the end I still made fairly basic mistakes because I did not know planning law back to front. However, I made a niche for myself in sustainability matters, which required learning only a few key sections of our planning rules, taking the greenest possible interpretation, and then making sure every development included the relevant eco-measures.

For example, Camden's 2006 Unitary Development Plan (the main pillar of the council's planning framework from 2006 to 2010) contained the sentence "We will require developers to conserve water." I translated that 'as all developments must install green roofs, rainwater harvesting and grey water recycling'. Officers were a bit surprised when this came up at every planning meeting for months, but eventually they and developers got the message and now it's standard unless there's a good reason why not.

A green roof is basically soil and plants instead of tiles or concrete. It provides fantastic natural insulation – warm in winter, cool in summer. It also slows down rainwater run-off during storms, which helps prevent flash floods. Green roofs become a haven for birds and other wildlife. They increase the lifetime of the roof itself because it's protected by the earth from damage. The plants and grasses also soak up carbon emissions. But best of all, when you look at the entire lifecycle of a green roof, it can actually be cheaper than traditional roofs. Green roofs are a fantastic example of how to rethink our urban landscape using sustainable methods. They're also good to look at!

As well as installing a green roof and rainwater harvesting and grey water recycling system, any new development in Camden has to produce 20 per cent of its energy from on-site renewables and make a contribution to sustainable urban drainage systems (SUDS).

Camden's new Local Development Framework, which was introduced in 2010, requires developers to offset any carbon produced by paying into a Community Infrastructure Fund, which is designed to be spent on creating a heat grid for the borough to speed up installation and connection of CHP systems. It also asks developers to aspire to the Passivhaus standard (see Chapter 5, page 94). It doesn't *require* it, just as Camden's previous Unitary Development Plan (UDP) didn't *require* a green roof or grey water recycling, but if councillors and residents make enough fuss about something, and can peg it on to some words in the planning rules and global or local events, then eventually it becomes the norm.

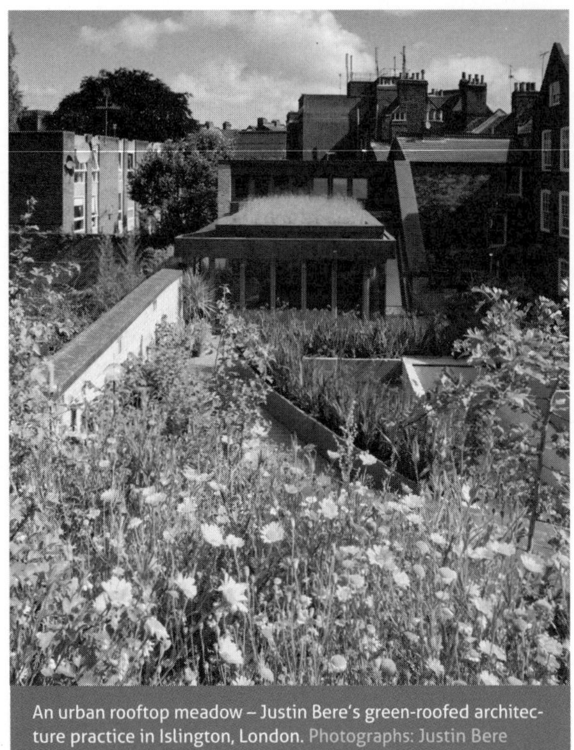

An urban rooftop meadow – Justin Bere's green-roofed architecture practice in Islington, London. Photographs: Justin Bere

PPS1 – an essential read

Everyone going down the path of teaching themselves planning law should read the supplement to Planning Policy Statement 1 (PPS1) on planning and climate change[1] (pictured opposite). It sets out a range of key objectives for planning in tackling the causes of climate change and providing for the unavoidable impacts. I have been told by civil servants at the Department of Communities and Local Government that they're surprised it's not used more often by community groups and councillors.

Dover District Council is the classic example of an authority that has read PPS1 and acted accordingly. Dover produced an evidence base to support the core strategy of its Local Development Framework (LDF) for sustainable construction and renewable energy policies in its area. This sets out local evidence to justify why Dover District should go beyond government targets and set tougher requirements for energy efficiency, renewable energy, water efficiency and other elements of sustainable construction.[2]

» *Scrutinise the wording of existing policies to see where language can be bent to your advantage, learn those key parts of the planning rules by heart, and use them ad infinitum*

How do you organise a campaign against (or in favour of) a planning application?

There's little or no point sending objection or promotion letters about a particular application to councillors who are on planning committee (Development Control or DC). Most simply put them straight into the recycling bin. But, somewhat contrarily, they do listen pretty carefully to objections made through the correct channels, and large campaigns definitely get the

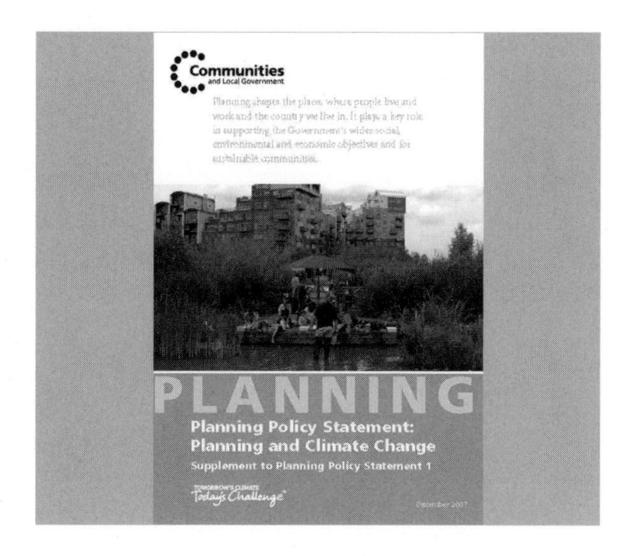

attention of planning committee members. In my experience, planning is one of the few places in a council where, if a community complains en masse and deputations speak cogently at committee, then they're highly likely to be given a sympathetic hearing.

Campaigning against a planning application

1. Work out what your planning reasons are for objecting to the application. You may need some help with this (see overleaf). Make a complaint on the council's website or send a letter to the committee clerk.

2. Contact your neighbours to seek support. Ask them all to send in letters of objection. One of the first things a councillor on planning committee does when he or she reads the officer's report is to see how many people have objected.

3. Contact your ward councillor to see if he or she will support you. With luck you will receive help in raising awareness in the local area. However, be aware that if your councillor is on the planning committee and expresses an opinion in public, he or she may subsequently be unable to take part in the decision-making process. Ask the councillor to request that the application be heard by the full committee rather than being decided by officers under delegated powers, which is what happens to 80 per cent of planning applications.

4. Contact the local papers and send them a photo of as many protesters as possible.

5. Persuade as many people as possible to attend the relevant planning meeting.

6. Ask for the right to speak at the planning committee meeting and choose the best speaker to address the committee. A dry speech by an expert will not help. Contributions from five or six people get messy as you're limited to five minutes in all. A lively speech by a good speaker who combines planning law with emotion is ideal.

Campaigning in favour of a planning application

1. If you want to promote something like a wind farm or a biogas facility, then it's worth trying to organise a Development Control Forum on the subject or a site visit for planning committee councillors. You'll have to organise this through the head of Development Control at your council.

2. You should try to head off objections before you reach planning committee by talking to as many potential objectors as possible. Understand what their complaint is and then try to circumnavigate it by reassurance. So, for example, you could also try to organise a trip to a site that has already successfully done what you want to do.

3. If the objectors won't engage with you, then you need to do a better job of marshalling your arguments both in written form and when speaking to the committee. Models and short clear Powerpoint presentations often help. A dull Powerpoint presentation is worse than bad.

Who can help you with planning issues?

Many community groups do develop expertise in planning matters – in my own area the Belsize Residents Association and Heath and Hampstead Society have members who understand planning law inside out and back to front. So do our Conservation Area Advisory Committees (CAACs), who usually do an excellent job of explaining the pros and cons of a particular scheme to planning committee members.

To find out which groups are thinking about planning issues in your area, simply get hold of an agenda for Development Control Committee (planning) and look through the objections. Sometimes you can piggyback these groups because your interests will coincide. But they will rarely be concerned primarily with climate change and natural resource depletion, so it pays to have someone or know someone who can concentrate on planning issues.

An expert from the industry

Sometimes a passionate local architect can be your answer, as they tend to have the ambition to do the right thing. I found one in Justin Bere of bere:architects when I was battling to get the Passivhaus standard introduced in Camden. Planning committees are convinced by a combination of passion and experience.

HOW THE CPRE'S KENT BRANCH INTERACTS WITH THE PLANNING SYSTEM

Sean Furey, Protect Kent, Campaign to Protect Rural England

The central area of our work is the planning process, which we engage with at various levels. Our national CPRE office lobbies central government and responds to national planning consultation issues (e.g. the new National Policy Statements and relevant draft legislation, e.g. the Planning Act 2008 when it was going through Parliament).

At the regional level we have an officer and a committee made up from the county branches who work to influence regional policy – in particular the Regional Spatial Strategies (RSS) (in our case the South East Plan).

At the county branch level we respond to Local Development Framework consultations and major planning issues (Kingsnorth power station, expansion of Lydd Airport, Kent International Gateway (KIG) freight depot, etc.). This can involve getting stuck in to public inquiries and we are currently at the KIG inquiry where we have given evidence, been cross-examined (not a nice experience) and we get the opportunity to cross-examine the applicants.

We also have district committees, made up of members from that area, and they review planning applications, respond to them or flag them up to the branch.

As well as working within the planning system, we also work outside it through public events, media and online campaigns and direct lobbying of MPs and councillors.

Confused? Our planning system is like quantum physics – if you think you understand it, you probably don't! I definitely don't.

Trained-up planning officers

In May 2006 there were no eco-specialists among Camden's planning officers. We had urban design and conservation experts galore – but nobody that knew anything much about sustainability. Every time an expert came to address the Sustainability Task Force I set up a follow-up seminar for our planners. Green roofs, grey-water recycling, Passivhaus – they started learning on the job, often in their lunch hours, which was very devoted of them. Couple that to the fact that half the planning committee were soon members of the Sustainability Task Force and you had an irresistible combination of officer education and political push. Planning officers now know to expect questions about sustainability issues and have adjusted their mindset accordingly. We didn't manage to make every planning application a paragon of eco-excellence, but we did make considerable progress.

Councillors who care about green issues

Once a rare breed, councillors who genuinely care about green issues, and who have the energy and the time to do something about it, are starting to become more common – thank goodness. Having so many members of my all-party Sustainability Task Force on planning committee in Camden during 2006-10 meant we were able to make good progress on the sustainability agenda. So seek out those eco-minded councillors and make friends with them.

» *Find or become a planning expert, because knowledge is power*

» *Offer free sustainability training to planners and councillors on planning committee – but only if you genuinely have the necessary expertise*

What should you push for in your Local Development Framework?

All planning authorities were required to formulate a new Local Development Framework (LDF) during 2009/10. It is supposed to make planning more spatial and strategic rather than piecemeal and site-specific.

This was, and may still be for some, a huge opportunity, since all councils have to consult the community several times before their LDF can be passed. But for many the process can be extremely frustrating.

> "We tried to engage with the Local Development Framework, but it was a failure. We got the head of planning to come along to an evening where he presented the LDF, and we presented the energy descent plan, where it was at, and the whole concept of peak oil and Transition because we wanted to influence the LDF. Basically he completely didn't get it. He was very nice about it, but my sense is that the council only does things on their own terms, so with something like an energy descent plan, they're not going to get it until the situation is upon us."
>
> Adrienne Campbell, Transition Lewes, East Sussex

Sean Furey of Protect Kent has been doing this stuff for a lot longer than any Transition Initiative. With time come successes, he says:

> "We now have good working relations with all the local authorities in Kent and many of the other public agencies, such as the Environment Agency and South East England Development Agency (SEEDA). We have had a successful judicial review against Kent County Council

and been influential in a number of planning application and planning policy decisions. Some policies in Local Development Frameworks are moving in the right direction but very slowly."

The following are some policies that community groups and councillors might push for in an LDF.

Energy

- **Reducing demand**. A clear commitment to reducing demand for energy via the planning system and analysis of how the planning system can help to meet carbon reduction targets.

- **Energy efficiency**. All public buildings and refurbishments of public buildings to use the Passivhaus standard for energy efficiency from now on so as to set an example; Passivhaus to be a requirement for all private-sector developments from 2013 and an aspiration until then; energy efficiency measures to be installed in all planning applications, especially new doors and windows; a look agreed for external insulation and double glazing, with a clear policy stating presumption in favour of energy-saving measures such as these.

- **Decentralised energy**. A clear sense of what your council's role should be in terms of negotiating CHP links between sites; a much better grasp of the risks of future fuel supply breakdown (peak oil, geopolitics, climate change); a discussion of the advantages/disadvantages of anaerobic digestion/energy from waste and a presumption in favour of small-scale units for reasons of resilience and localism rather than large out-of-area facilities.

- **On-site renewables**. At least 20 per cent on-site renewables – but be flexible if developers a) try for Passivhaus and narrowly fail, or b) suggest an off-site renewables approach that can be shown to be better in terms of additionality (i.e. wouldn't have existed otherwise) and more 'bang per buck' than on-site renewables; a special Supplementary Planning Guidance on which renewables make sense for an area.

Water

- Waterscapes to be built using rainwater wherever possible.

- 'Lost' (culverted) urban rivers to be opened up.

- Drinking-water fountains to be added in the public domain on all major developments.

Transport

- Car club spaces to be installed on or near major developments, and all residents in urban areas to be within a few minutes' walk of car club spaces, thereby obviating the need for private cars for all except the disabled.

- Contributions to cycling infrastructure (separated cycle lanes or cyclist priority roads) and more secure cycle storage to be required.

Green spaces

- Commitment made to an increase in green space annually and per development site because of carbon sequestration, the urban heat island effect, storm-water attenuation, food growing, urban amenity, mental health, etc.

- All sites to increase the amount of green space compared with pre-application.
- Green space potential to be mapped and included in LDF as aspiration/target.

Food

- All major developments to provide food-growing space either at ground level, vertically (on walls) or on roofs.
- Sites unused for more than six months to be turned over to the local community for food growing until developed.
- Major developments to be encouraged to provide sites for farmers' markets.

Biodiversity

- All sites to enhance biodiversity, and there must be some form of post-installation measurement.
- Carbon-sequestration value of trees to be included in policy.
- New tree policy of 'one out, two in'.

Health

- LDF to be much tougher on air quality, especially with regard to inefficient diesel engines, e.g. in construction-management plans.
- All developments to be required to design for health. i.e. prove that they have minimised, or better still eradicated, the use of toxins in building materials.
- Maximisation of natural light wherever possible to be included in policy.

Demolition

- A presumption that buildings be refurbished, preferably using the Passivhaus standard, rather than demolished, even when they are not in a Conservation Area or are not listed.
- Onus to be put on developer to prove that demolition is the more sustainable option.

General

- Clear acknowledgement that sustainability requirements are to apply to all applications, not just those of five units or more.
- The need for increased density and population in cities to be disputed, citing long-term unsustainability given peak oil, etc.

Don't worry if your Local Development Framework has been agreed. Just turn over the page metaphorically and start the process of revising it.

Is carbon offsetting ever acceptable?

Carbon offsetting is highly controversial among environmentalists, and rightly so. The price of carbon used in the offsetting business is minimal, because the market price of carbon is far too low, and investing in trees on the other side of the world, trees that might be chopped down one day, can hardly be a very sensible way to deal with the carbon problem.

"The public must be firm and unwavering in demanding 'no offsets', because this sort of monkey business is exactly the sort of thing that politicians love and will try to keep. Offsets are like the indulgences that were sold by the church in the Middle Ages. People of means loved indulgences because they could practise any hanky-panky or worse and then simply purchase an indulgence to avoid punishment for their sins. Bishops loved them too, because they brought in lots of moola. Anybody who argues for offsets today is either a sinner who wants to pretend he or she has done adequate penance or a bishop collecting moola."

James Hansen, NASA scientist[3]

That said, there are, I think, two perfectly legitimate ways that individuals and local authorities can carbon offset.

1. Milton Keynes Council has introduced an innovative carbon offset fund for new developments. Anyone seeking planning permission has to make their new building as energy-efficient as possible and then offset into the Milton Keynes Carbon Offset Fund any remaining carbon expected to be generated over the lifetime of the site. The council will then spend that money on measures that reduce carbon emissions, e.g. insulating private homes or council houses. I call that ingenious.

2. Eastleigh added a voluntary component to the planning compulsion idea to give local businesses already offsetting into dubious schemes on the other side of the world the chance to invest in CO_2 reductions locally.

In my opinion, if you keep it local and verifiable there is life in carbon offsetting. The one problem is the price of carbon. Forget the market price – use the actual cost to a local authority of extracting a tonne of carbon from the atmosphere.

In early 2010 Brighton Council was looking to set the bar extremely high by proposing a carbon price of £1 per kilo of carbon or £1,000 per tonne. That was based on the costs of retrofitting housing with various measures, from insulation to solar thermal, to reduce emissions by 1 tonne per year.[4] That should give developers a serious incentive to do the maximum possible on energy efficiency and renewables before they think about offsetting. Bravo Brighton!

» *Ask your council's Planning Department to introduce a local carbon offset fund and set a high price of carbon*

What does positive collaboration between a council and the community on planning look like?

The London Borough of Sutton is the site of BedZED, the UK's largest low-carbon community. Some mistakes were made in the building of BedZED, which is hardly surprising given how much new ground the architects were attempting to break, but it has nevertheless served as a beacon of eco-living and has inspired further action. Sutton is now working with the community to turn the surrounding area – Hackbridge – into the UK's first truly sustainable suburb.

In 2009 the council unveiled a draft masterplan for the Hackbridge area, including proposals for new eco-friendly homes, more shops, leisure and community facilities, jobs, sustainable transport and pedestrian/ cycle initiatives, improved networks and open spaces. Just over 90 per cent of respondents to the subsequent consultation supported the proposed vision to make Hackbridge the UK's greenest suburb, using the principles of One Planet Living (see Chapter 1, page 30).

The council is also promoting retrofit initiatives for existing homes and the highest standards of sustainable design and construction for new development in the area, which will connect to local renewable heat and energy sources, with a view to engineering significant reductions in carbon emissions. The results will surely provide more inspiration for communities and councils around the country.

A vision of the future? BedZED, Britain's biggest low-carbon community, has triggered moves by Sutton Council to green the whole of the surrounding area. Photograph: BioRegional

Chapter 9

PROCUREMENT

Questions answered in this chapter:

- How can community groups and councillors penetrate council procurement processes?
- Are EU trade rules a help or a hindrance when trying to procure sustainably?
- Can you ask for community benefits or specify eco-products in contracts?
- How can councils use their buying power to start a carbon dialogue with suppliers?

Paper or electronic money has meaning only as long as people believe it will buy something. Another way of phrasing it is to say that it doesn't exist without trust. However, trust is in short supply in a global financial system built on illusion and debt. Money is no longer backed by anything real – it is backed only by a slightly insane belief that the (Ponzi) pyramid will go on growing and that there will always be more money – created by private banks or government – to keep myriad asset bubbles up in the air.[1] At some point I expect the illusion of money to fail.[2]

Loving the Brixton Pound - Mayor Wellbelove of Lambeth shopping with the first urban local currency in the UK. Photograph: Transition Town Brixton

I was therefore going to write a chapter about local currencies, Local Exchange Trading Schemes, time banks, credit unions and the like, because these mechanisms feel like a key part of the solution to the possible collapse of the financial system. Some forward-looking councils are starting to work on parts of this agenda with local community groups. For example, the London Borough of Lambeth has been very supportive of the Brixton Pound, the London Borough of Islington has given its local credit union a lot of help, and the London Borough of Camden funded the creation of the Holy Cross Time Bank[3] as a way to help those recovering from mental illness back into productive society.

Essentially there are two parts to a coherent solution to the financial crisis – increased relocalisation and resilience by communities on the one hand, which is a theme running right through this book, and local systems of exchange on the other. But the definitive book has already been written on this subject: it's called *Local Money: How to make it happen in your community* and it's by Peter North,[4] so I recommend you read that if you want to know more about alternative systems of exchange. I will instead concentrate on the not-inconsequential subject of council procurement.

How can community groups and councillors penetrate council procurement processes?

Britain's biggest council, Birmingham, spent £3.38bn in 2009/10 on goods and services.[5] Local government as a whole in the UK was projected to spend £173bn in 2010.[6] These are big sums. Sustainable procurement – which minimises damage to the environment and health and maximises social and economic benefits – is therefore one of the principal drivers a council has for addressing climate change. In fact it's probably the biggest lever, but few councils use it systematically and it's hard for community groups to influence the process of procurement unless the thing being bought is big and visible like a new school. That's why purchasing is one of two areas of local authority work, along with planning, where I think councillors and community groups need some help.

These days those who espouse sustainable procurement are getting a better hearing in most quarters, but it's nevertheless hard work to help policymakers and other non-expert stakeholders to understand the key concepts and to get them mainstreamed. The important thing, for community groups and councillors, is to understand what's going on and to ask the right questions.

Employing experts

If you want to achieve anything on this topic, then you have to have sustainable procurement experts on the inside. In 2006 Camden was rated one of the top five public authorities in Europe for sustainable purchasing, but shortly afterwards the council scrapped its sustainable procurement specialist, the person who'd done so much to make it a pacesetter on this agenda. Within two years a lot of expertise that had been built up was lost and Camden had to reinvent the wheel.

However, even if you do manage to get a sustainable procurement expert employed, he or she has to be able to play the system and know how to deal with the politics of executive management.

> "You can have the best tools and toolkit and wise words, but if the person who's driving it, the sustainable procurement expert, if they do not have an understanding of change management, then forget it. You can have the best sustainable procurement expert in the world, but if they don't know how to interface and re-communicate their knowledge and their pressures, within their language at the appropriate time and in appropriate ways, it will get nowhere.
>
> There needs to be a Holy Trinity. You've got your sustainability procurement expert – the geek who knows the policies. There has to be a director-level champion. And then you need to have an informed elected member. If you've got one of those three things missing, the impact is going to be short-term and highly limited."
>
> Fay Blair, sustainable procurement expert, Local to Global

Stand-alone sustainable procurement policies

When I first started looking at this subject I thought that having a stand-alone policy based on best practice around the country was self-evidently a good thing. According to 'susproc' experts, however, the jury is still out on this. Fay Blair again:

> "You want procurement policies and protocols that have embedded sustainable procurements, threads, facets, within that. If you separate that out as a distinct and separate entity, you run the risk, as I've seen and witnessed in many councils, that the hard-core procurement professionals will read the mainstream documents first and then as an annex and ancillary pay heed to whatever subsidiary documents or documentation on sustainable procurement might co-exist or come under that."

If you're building a school or a care home, then the worst mistake you can make is to add on the sustainability measures, like photovoltaic panels, at the end. If you do that, then as soon as the budget comes under pressure they can be top-sliced out. Sustainable procurement is the same. If you separate out sustainability, then there's a chance it'll become an add-on that can be easily removed.

Best practice

Fay says there are departments or sub-departments that are doing the right thing, but that all too often what they're doing is not widely known about and not systematically or coherently applied across their own organisation. However, sometimes good practice can be implicit rather than explicit, she adds.

> "When I interviewed one council in the North East that was one of the top two performing Corporate Performance Assessment authorities [as measured by the Audit Commission until 2009], what was very striking, when I had that round-the-table set of interviews, was that because of the competency and the maturity of understanding of senior professionals, they had a lot of these things embedded in the whole warp and weft of what they were doing, without using any of the usual sustainability tags, jargon or nomenclature. In other words, if you asked 'Do you have a sustainable procurement policy?' they would say 'Well, no, we don't,' but when you read their policy, my goodness, they had everything in there."

Fay helps to run the National Sustainable Procurement Stakeholders Group, a voluntary body that seeks to promote sustainable procurement among local authorities and other regional bodies. It has a best practice website called the Sustainable Procurement Cupboard (www.procurementcupboard.org), aimed at procurement professionals looking for case studies, tools, primary documents and contacts to help them deliver on multiple public sector targets (see box, opposite). With a little effort, it can be used by councillors and community groups as a tool to help them to ask the right questions.

There's also a good guide published jointly in 2007 by the Improvement & Development Agency, the Local Government Association and the North East Centre of Excellence called 'Sustainability and Local Government Procurement'.[7] It draws on the experience of English and Welsh local authorities and provides practical advice on how a commitment to sustainable development can be turned into an effective procurement policy and built into processes and strategy.

Ten actions to embed sustainability legally into the procurement process

- Understand your powers to incorporate sustainability initiatives into your procurement.
- Use available guidance and legislation to justify your actions.
- Embed sustainability into your organisation's policy framework and document it.
- Include sustainability at the earliest possible stage in the procurement process, i.e. when making the business case for the procurement, so that it is within your core requirements for the procurement.
- Ascertain technical capacity and ability to deliver economic, social and environmental well-being outcomes within the contract requirement by asking relevant questions in Pre-Qualification Questionnaires.
- Describe sustainability requirements in your specification in as specific a way as possible.
- Get the bidders to expand in their method statements on their delivery of sustainability-based outcomes and outputs described in the specifications with sufficient clarity that you will be able to measure the contractor's performance.
- Use contract conditions to enforce social and environmental requirements in contracts.
- Develop criteria for awarding contracts which address how value for money will be measured against comprehensive specifications which embed social and environmental issues as part of the whole contract requirement.
- Educate and train your people so that sustainability becomes the norm in procurement.

Sustainable Procurement Cupboard[8]

» *Ask your council's procurement team to send a representative to the National Sustainable Procurement Stakeholders Group*

» *Try cutting and pasting the list in the box (left) into a letter to your local councillor and ask how the council matches up!*

Are EU trade rules a help or a hindrance when trying to procure sustainably?

In the past, EU single market rules have proved to be the biggest obstacle to more sustainable purchasing. Now thinking at the EU level is moving in the direction of those of us who want to buy green.

As part of its six-month EU presidency in 2008, the Slovenian government hosted a conference on sustainable consumption and production. One of the key recommendations was an EU Directive on Green Public Procurement (GPP) to make responsible purchasing mandatory for all public authorities. By responsible it meant low-carbon, fair trade, and from sustainable resources. The directive was never introduced but there was a general agreement among EU nations that they should all be moving in this direction.[9]

There's now an EU Green Procurement Training Kit, which can be downloaded from the European Commission website.[10] The toolkit consists of three independent modules, each designed to overcome a specific problem identified as a barrier to the uptake of Green Public Procurement within a public organisation:

- A strategic module, which seeks to raise the political support for green public procurement within an organisation, targeting in particular decision-makers.

- A legal module, which seeks to clarify legal issues and is designed for both strategic and operational levels.
- An operational module, aimed at purchasing officers responsible for the preparation of tender documents, which includes concrete examples of environmental criteria for 11 product and service groups.

Can you ask for community benefits or specify eco-products in contracts?

Councils spend millions on food for schools, care homes and their own canteens. Under EU trade rules a council cannot specify local or organic or seasonal food in a contract, but it can specify fresh food, which is one way to achieve a somewhat more sustainable food supply as fresh food tends to be more local and less processed. There's also European and UK legislation that permits a 'community benefit' approach in public procurement. The European Public Procurement Act 2004 allows community benefit in procurement on a permissive basis, and Section 3 of the 2000 Local Government Act gives councils the right to act under power of community well-being. Both of these could allow circumnavigation of EU trade rules, but for a council procurement team to take the risk would require some pretty significant managerial and political buy-in.

In Europe the French and Italians have been running rings around EU trade rules for years by specifying hundreds of foods as having a unique link to a geographical place and/or culture. Protected Denominations of Origin (PDO), Protected Geographical Indications of supply (PGI) or Traditional Speciality Guarantee (TSG) are all perfectly acceptable ways to give food protected name status under EU law.

A council in Tuscany or Burgundy could probably guarantee that all its food purchases were kept local by specifying regional specialities such as Prosciutto Toscano (Tuscan ham) or Volailles de Bourgogne (Burgundy chicken). In Britain that's not an option; in 2006 there were only 31 products in the UK registered by area or speciality, including Jersey potatoes, Scotch Beef and Blue Stilton.

In future the key for those working to secure better, more sustainable food for schools and other council services in the UK is likely to be evidence-based.

> "If the whole idea of quality food, nutritious food, is something your parent-teacher-governor base is saying is important, then you've got the legitimate mandate to specify that. What stakeholders, users of services, think about what they want, what is important to them – my understanding is that the legislation and the protocols that local government and the public services are expected to track and evidence and follow more robustly, is taking their views into account."
>
> Fay Blair, sustainable procurement expert, Local to Global

» *Find out when your local school food contract is next up for grabs, then ask the council to consult parents and teachers and to use the results as an evidence base for the contract, and organise a campaign at the school to lobby the council for the right to be consulted*

Helping small businesses to win contracts

Small local businesses, which employ far more people than large businesses and which spend far more of their earnings locally, are usually frozen out of council contracts when price and size are the defining characteristics of the tender. However, there are several ways to help them.

1. Reduce the value of tenders. The smaller the tender the less interesting it is likely to be for the big boys.

2. The council can help small businesses to fill in tender replies, which usually take a lot more time than an overworked small-business owner has to spare. Of course, a council cannot fill in a specific tender reply for a small business, but it can run workshops to help small businesses with upcoming tenders.

3. Councils can use the locally gathered evidence base to ask for community benefits from tenders. None of this means a tender can be judged on how small or local a company is, but it does make the playing field a bit flatter.

» *If you're a small-business owner, ask for help on filling in tender applications from the Procurement Department of your council*

Helsinki's gas-fired buses

Helsinki City Council put out a tender for gas-fired buses in 1997 with the explicit aim of reducing carbon emissions, improving air quality and reducing noise pollution. However, one company complained that meeting environmental criteria brought no economic advantage to the authority and was discriminatory towards companies who did not supply gas-engined buses.

In a landmark ruling the court decided that:

- Environmental criteria were linked to the subject matter of the contract and that the council's organisational goals included reduction of carbon emissions, noise pollution and noxious emissions.
- The environmental criteria were adequately specific and measurable, and actually verifiable (scientific).
- Councils procure for the benefit of citizens not procurement departments.

The importance of this last point is hard to underestimate. Councils are not businesses. Businesses buy for the benefit of their business. Councils buy for benefit of their residents.

See http://www.ens-newswire.com/ens/sep2002/2002-09-18-03.html

Eco-products

Under EU trade rules you cannot judge tender responses on how environmentally conscious a company is, but you can specify eco-goods, and this is what you should do as far as possible. So, for example, if you ask the market for paper, then you can't judge the tender on whether respondents have offered recycled paper. But you can ask for recycled paper!

It helps if your organisational goals include environmental and sustainability issues. My favourite sustainable procurement case study is that of Helsinki City Council and its gas-fired buses (see box), which established the principle that a council is different in purchasing terms from a business because it procures for the benefit of citizens not the business.

» *Ask your ward councillor to ask the council's procurement department if it is systematically specifying eco-products wherever possible*

Lowest price is . . . lowest price

Frankly, no local authority should be awarding contracts on the basis of lowest price any more. EU law allows value-for-money criteria to be used, which is defined as the optimum combination of whole-life (or life-cycle) costs and quality to meet the user's requirement.

The upfront cost of a green product may appear more expensive, but when all the costs (throughout the working life of the product) are analysed, overall the greener alternative may well prove to be cheaper over time. If contracting authorities wish to ascertain which products are most cost-effective for them they need to apply Life-Cycle Costing (LCC) approaches in their procurement decisions. This means comparing not just the initial purchase price of a product but all future costs as well, e.g. usage costs (energy/water consumption, consumables such as ink or paper), maintenance costs and disposal costs or resale value.

Take the Passivhaus energy efficiency standard, for example. On average a Passivhaus building costs 0-7 per cent more than conventional build, but if you include lifetime energy bills in the build costs, then it costs less than conventional build.

Understanding whole life costs and CO_2 in a combined tool

The sustainable development charity Forum for the Future and Fife Council have developed a tool that helps procurers to understand the CO_2 cost as well as the whole-life financial costs of a product: the *Whole life costing (+CO_2) – user guide.*[11] The tool enables you to calculate how much CO_2 will be emitted by different products during the contract period, and how much these emissions will cost under the CRC Energy Efficiency Scheme (see Chapter 2, page 42) or other carbon-pricing schemes. In October 2009 Fife Council was awarded the UK Government Opportunities Sustainability Award for its use of the tool.

» *Ask your local councillor to ask officers how many contracts are being specified primarily on lowest cost, and how often lifetime energy bills are being included in the contract specifications*

How can councils use their buying power to start a carbon dialogue with suppliers?

CAESER (www.caeser.org) is a private company that tries to help organisations to ask the right questions of their suppliers so as to maximise sustainability outcomes. CAESER will send out a survey to suppliers about environmental, social and economic sustainability then score the responses.

Good public-sector bodies are probably already doing this through supplier engagement programmes, but most local authorities are not. It feels like a helpful first pass for those who are facing a steep learning curve, albeit an expensive one.

The Carbon Disclosure Project (CDP – www.cdproject.net) is a global charity that does something similar (at much lower cost) but concentrates on carbon and, since 2010, water. CDP began as a way to help institutional investors to assess their carbon risk (Investor CDP) and became a supply chain tool when Wal-Mart understood that it could use it to cut out waste (CDP Supply Chain). In 2008 CDP Public Procurement was created after Camden had trialled the concept the previous year (see box, opposite).

» *Encourage your council's Procurement Department (via your ward councillor if you're a community group) to put carbon disclosure in all contracts*

Camden and carbon disclosure

I had been looking for ways to leverage Camden's buying power for the benefit of the planet because Camden spends a staggering £500m a year on goods and services. When I discovered CDP it suddenly occurred to me that I could use carbon disclosure to start a dialogue with suppliers about climate change and begin the process of decarbonising our supply chain. So I persuaded Camden Council to become the first public authority to sign up to the CDP process.

At this point I was primarily thinking that raising awareness among our suppliers was a worthy goal in itself, and that, although hard to quantify, asking suppliers to disclose their greenhouse gas emissions and any strategy they might have for reducing them was a legitimate policy tool for assisting with the general emissions reduction strategy of Camden Council. However, my secondary thought was about how to use carbon disclosure to improve our sustainable procurement policy without falling foul of EU trade rules.

Shortly after Camden joined the CDP process as a guinea pig the UK government issued its National Indicator set (see Chapter 2) and DEFRA issued guidance recommending that councils put carbon disclosure in new contracts as a way to gather data for the new National Indicator 185 – reducing the carbon emissions of a local authority.

NI 185 requires councils to report their direct emissions as well as, crucially, those of their outsourced services. Most of the services Camden delivers – e.g. rubbish collection, recycling, housing repairs, highways maintenance, upkeep of parks – are in fact outsourced. Gathering outsourced emissions data for NI 185 is a headache. Suppliers need to agree to do it, they need to know how to do it, and they need to be able to apportion their emissions by customer. But the DEFRA guidance at least gave local authorities the right to ask for carbon disclosure in contracts. In time I think this will be standard practice.

Chapter 10

RECYCLING

Questions answered in this chapter:

- How can we do more to stop waste creation?
- Can Pay As You Throw or reward or compulsion help to increase recycling?
- How can we persuade people to keep using stuff until it's worn out?
- How can we get residents to fix broken stuff rather than buy more?
- How can we ensure our recycling is good for the planet and that we're doing as much as possible?
- How can we maximise the extraction of energy from our food waste?

If you ask most people what they're doing to prevent dangerous or runaway climate change, they'll usually tell you they recycle or they might say they refuse plastic bags in shops. On a bad day this fills me with despair because there are so many other issues that are more critical, such as flying, eating meat, driving and accumulating stuff. On a good day I tell myself it's a start, an opportunity even.

When the Chair of the Hampstead Women's Club asked me to give a talk to her fairly well-heeled members about how to improve their recycling techniques I took a deep breath and said I could, but that I would feel obliged to talk about the real problems as well. After some back and forth negotiations it was agreed – I could also talk about materialism and One Planet Living (see Chapter 1, page 30) as well as which plastics could be recycled. One Hampstead Women's Club member even invited me into her home to do an eco-audit of her family's lifestyle to illustrate the issues to the rest of the Club.

I was so nervous on the day of the talk that I cut myself shaving for the first time in years. But I needn't have worried. They listened, they were shocked and they agreed that changes needed to be made. Some of them even came up to me at the end and suggested forming a Transition Hampstead group, which was beyond my wildest dreams. So maybe the lesson here is take every opportunity to reach a new group or network and to use a non-confrontational but firm style that stresses the problems as well as the potential positives of life after oil.

Most people now know the Reduce-Reuse-Recycle mantra. In this chapter I have expanded the three Rs to five, by adding Repair and Recover (as in recover energy or compost).

Reduce – How can we do more to stop waste creation?

There are some really obvious ways in which the amount of waste, in particular certain types of waste, that routinely goes to landfill can be reduced.

Plastic bag bans

Plastic bags are both nothing and everything – they are a tiny part of the global problem but a very potent symbol of our wasteful economic system and, where they come into contact with nature, they can be highly destructive.

The world uses more than 1.2 trillion plastic bags a year – an average of about 300 bags for every adult, or 1 million bags used per minute. UK citizens are estimated to use 290 plastic bags per person per year. On average we use each plastic bag for 12 minutes before discarding it. It then can last in the environment for decades.[1]

Nearly half of windborne litter escaping from landfills is plastic (47 per cent) – much of it plastic bags. About 80 per cent of all marine rubbish comes from off the land, and nearly 90 per cent of that is plastic. In June 2006 the United Nations Environment Programme estimated that there was an average of 46,000 pieces of

How to start a plastic bag ban in your community

On May 1st 2007 Modbury, a small market town in Devon, became the first place in the British Isles to stop issuing plastic shopping bags when goods are purchased. The move was spearheaded by Rebecca Hosking, the BBC wildlife camerawoman whose documentary *Message in the Waves* about how seaborne plastic was decimating wildlife in Hawaii inspired her to act. Here's a summary of how she and the local community think they managed to persuade Modbury to go bag-free.

- Be prepared to fight the fight without help from others if necessary (at least at the start).
- Know the issues in your community if you want to gain the trust of traders.
- Do your homework on what plastic is doing to the environment, to animals and to humans.
- Don't get bogged down in potential grants and red tape.
- Approach traders directly, not via flyers or emails and be prepared to answer their questions.
- Invite traders to a screening of *Message in the Waves.*
- Visit those who didn't go to the screening.

More than 1.2 trillion (1,200,000,000,000) plastic bags are used around the world every year. Photograph: iStock

- Use traders who are already on board to persuade others.
- Ask traders what they would want from a non-disposable bag.
- Research every type of environmental bag on the market and commission one you feel comfortable with.
- Create a 'contract' for traders to sign saying that they will go plastic-bag-free.
- Set a date for the town to go plastic-bag-free.
- Go for it!

plastic debris floating on or near the surface of every square mile of ocean.[2]

> "Plastic bag litter is lethal in the marine environment, killing at least 100,000 birds, whales, seals and turtles every year. After an animal is killed by plastic bags, its body decomposes and the plastic is released back into the environment, where it can kill again. Plastic bags do not biodegrade, they photodegrade – break down into smaller and smaller bits, contaminating soil, waterways and oceans, and entering the food chain when ingested by animals."
>
> Rebecca Hosking, wildlife film maker and founder of the Modbury plastic bag ban

If you can persuade people to cut down on plastic bag use, maybe, just maybe, they will think about how they could apply the same principles to the rest of their lives.

Packaging laws

In Germany shoppers have the right to leave any excess packaging at the till and retailers are legally obliged to recycle it. Unfortunately there's no such law in the UK. But we do have packaging laws that brave authorities such as Northampton have used to good effect.

Trading Standards has the power to regulate under the Packaging (Essential Requirements) Regulations of 1997 and 2003, and can therefore take to court any company that over-packages. Annex II of both documents reads: "Packaging shall be so manufactured that the packaging volume and weight be limited to the minimum adequate amount to maintain the necessary level of safety, hygiene and acceptance for the packed product and for the consumer." In 2007 the new Producer Responsibility Obligations (Packaging Waste) Regulations came into force, which incorporated the 1997, 2003 and 2007 Packaging Regulations and which added a lot of technical detail to the previous Acts.

Northampton, Cambridgeshire, Oldham and packaging laws

Four firms have been successfully prosecuted since the packaging regulations were introduced in 1999.

- Office World was found by Northamptonshire trading standard officers to consistently use large boxes to dispatch very small items. In one case, the company used a box over 14 times larger than the goods it contained. The firm was fined and ordered to pay costs.
- Northamptonshire officers also prosecuted a butcher who sold pre-packed meat on an upturned polystyrene tray inside another larger tray. He was fined.
- In Cambridgeshire a customer complained to trading standards after buying some Cadbury's Giant Chocolate Fingers at Tesco which were not actually giant. They came in a tin over 16cm long but inside were individually wrapped chocolate fingers less than 12cm long. The company responsible, Burton's Foods, was fined.
- Officers in Oldham took a firm called Nadia Luciani to court after discovering a tin of dried mushroom powder had a false bottom, making it three inches longer than necessary. The company was fined.[3]

Launching an over-packaging campaign would be an excellent way for local authorities to tap into the energy, enthusiasm and concern of residents by asking them to notify the council of blatant examples of over-packaging. Trading Standards staff could then engage the manufacturer in question and, if no change is forthcoming, the council could consider legal action. This doesn't have to be a campaign against all manufacturers. A few highly publicised test cases might help to change mindsets. Equally, community groups could lead the campaigning and then ask the council's trading standards team to take action. But if you do this, then someone in the council – an officer or a politician – needs to be supportive of the risk and cost of prosecution, so you'll need to do some lobbying first.

Transition Belsize persuaded its local Budgens supermarket to stop giving away plastic bags and to introduce recycling bins at cash tills for excess packaging in an attempt to replicate what happens in Germany.

Budgens and Transition Belsize

I'm not sure at what point Andrew Thornton, the owner of the Belsize Park Budgens franchise, 'got it', but as soon as he did, dramatic change became possible at his supermarket. As a result of working with Transition Belsize his stores now:

- sell over 1,500 local food lines (sourced within 100 miles of the store) out of 10,000
- send their food waste to be turned into biogas and then electricity
- put only 6 per cent of total waste into landfill
- have recycling bins at the till for excess packaging
- offer recycling facilities for excess packaging, mobile phones and printer cartridges
- cover their fridges and freezers at night to save energy
- have a sustainable fish policy using a 'traffic light' coding system
- use 70 per cent fewer plastic bags than in 2008 and have introduced a charge of 2p for 'single use' carrier bags to reduce usage further
- offer bicycle deliveries to those who live within two miles of the store, both for food if they spend more than £25 in his shops and for moving general stuff around, for a £5 charge.

In addition, Andrew has always said that he will sell any food that Transition Belsize can produce and will help with packaging it if need be. At the 2009 Belsize Green Fair he supported the making of a cob pizza oven on his forecourt. He's also been pushing for Camden to introduce a local pound, which will surely be a key factor in the currency's eventual success if it's ever created.

Andrew may be unique in the supermarket world, both because he understands the problems and because he has the power to do something. He has to buy 90 per cent of his produce from the distribution company behind Budgens, but the rest he can buy where he likes. He's also reasonably free to do what he wants with his stores physically. By contrast, although Transition Belsize developed a good relationship with the manager of the local Tesco, and even persuaded him to come to a screening of the BBC's *A Farm for the Future* programme, he was unable to do much except secure a little funding for Transition Belsize projects, and do some work on plastic bag reduction during Belsize Eco Week.

Cooking pizza with cob – Andrew Thornton looks on as his Deli Counter Manager Marco prepares to make pizzas in a wood-fired oven made of mud and straw. Photograph: Sarah Nicholl

» *Pursue a two-pronged campaign of positive engagement with your local supermarkets and lobbying of the council's Trading Standards Officers to get them to use the Packaging Regulations*

Introducing a Zero Waste to Landfill Strategy

The goal of Zero Waste is to minimise and ultimately eliminate waste. It is the expression of the desire in some countries, regions or cities to move away from

Going Zero Waste in Bath

Bath's progress towards Zero Waste began with a collection of old newspapers by Avon Friends of the Earth in the 1980s. In 1991 the council employed Avon Friends of the Earth, by then a not-for-profit social enterprise, to do the collections. Within a few years it was collecting paper, bottles, cans, aluminium foil and old clothes – occasionally in a horse and cart!

By 1994 Bath was recycling a remarkable 25 per cent of its waste whilst recycling was barely on the agenda of most councils. Plastic recycling was added that year, a decade ahead of most local authorities in the UK. Less happily, Avon Friends of the Earth went bust, but its place was taken by another social enterprise, ECT Recycling. It has been argued that the involvement of community groups/social enterprises, plus an administration that understood the value of recycling, explains ECT Recycling's success in Bath.[4] In many ways I think this is the perfect example of how the community and the council working together can do the right thing for the planet. The community came up with the idea. The council helped to underpin the idea by commissioning the community to do more of it.

By 2003 the recycling rate in the new unitary authority of Bath and North East Somerset had reached 33 per cent despite its merger in 1996 with Wansdyke Council – which had a recycling rate of just 2 per cent. That merger slowed but did not stall progress towards the Zero Waste target and by 2008 the recycling rate was 43 per cent.

When I talked to the Head of Waste and Recycling Strategy at Bath and North East Somerset in 2007 no work had been done to quantify the effects of declaring a Zero Waste Strategy. However, she was convinced that the policy had had beneficial effects, not least in conveying the waste minimisation message to residents. Residents, in turn, had reacted favourably to the policy in feedback to the council, which often holds Zero Waste events, producing still more positive publicity.

Challenges to Establishing a Zero Waste Strategy, according to Bath & NE Somerset Council[5]

- 'Zero' is a target to encourage innovation - not an 'absolute'.
- Currently there is no legislative requirement to reduce waste.
- Waste-generation processes involve all sectors of society and are not easily addressed at a local level.
- Unknown future – future waste minimisation processes and technology will change.
- Reaching beyond – it is relatively easy to plan to achieve 50 per cent recycling & recovery. Innovation is needed to get beyond these levels.
- Sustaining and building momentum. We need to continue moving beyond our initial enthusiasm and success.
- Sustainable solutions – we will need a new way of thinking at a national level to achieve reduced consumerism and increased environmental stewardship at a local level.
- Resources – investment in new technology and process will be required to achieve progress.

the linear model of resource use, where resources are extracted, turned into goods, consumed and discarded. A Zero Waste Strategy seeks to: reduce consumption; ensure that products are made to be reused, repaired, recycled or composted; minimise residual waste; and optimise recycling.

New Zealand is often credited with being the first country to declare a Zero Waste goal. The policy has also been embraced by several major cities such as San Francisco, which has set a goal of Zero Waste to Landfill by 2020. In 2001 Bath and North East Somerset became the first local authority in the UK to commit to a Zero Waste Strategy as a policy framework to test policy options against and as a way to help its residents, institutions and businesses focus on the need to reduce the amount of waste created (see box, opposite). The target date set was 2020, with the interim goal of 50 per cent recycling of household waste by 2010, as well as an explicit aim of arresting growth in waste, despite the predicted population increase in the area. A number of UK local authorities have followed Bath down the Zero Waste route and in 2010 the Scottish government announced that it was going Zero Waste.

Recycle – can Pay As You Throw or reward or compulsion help to increase recycling?

Under the Waste Minimisation Act 1998 "a relevant authority may do, or arrange for the doing of, or contribute towards the expenses of the doing of, anything which in its opinion is necessary or expedient for the purpose of minimising the quantities of controlled waste, or controlled waste of any description, generated in its area."

Nappy Happy in North London

It's been my privilege on several occasions to act as compère at the Islington-Camden Fashion Show for babies and their reusable or 'real' nappies. Toddlers parade up and down (or off!) a mini-catwalk wearing the latest brightly coloured reusable nappies while I announce their names and why they (or their parents!) think real nappies are a good thing. It's definitely one of my favourite events of the year.

By the time a baby is potty-trained most parents will have used between 4,000 and 6,000 disposable nappies, most of which end up in landfill sites, creating methane. This has to be because the parents don't know that modern eco-friendly nappies are not the terry towelling of yesteryear. These days reusables are made of layers of muslin that are as absorbent as a disposable nappy. What's more they don't leak, they don't cause nappy rash (or at least no more than disposables), they're good for the environment and they're a lot cheaper.

If you home-wash them costs can be as low as £1 a week. Or, for the price of disposable nappies, you can use a laundry service that collects used nappies and provide a fresh supply. And many councils even give you money towards real nappies.

» *Ask your council's recycling team if it subsidises real nappies (see box above) and if it doesn't, ask your councillor or the relevant Executive Member to consider doing so*

Pay As You Throw

The government's 2007 Waste Strategy recommended introducing 'Pay As You Throw', which means charging residents according to how much waste they create. Under one version of the scheme, which was tested in south Norfolk from 2002 to 2008, a microchip in householders' bins was used to weigh their waste. They were then charged accordingly. In principle I think this is a good idea, although the bin tax was mercilessly castigated by the *Daily Mail* and others, and South Norfolk council eventually abandoned the pilot. The government promised more trials. They never happened, although in March 2010 it was reported that 68 councils were using 'chip bins' but without actually using the embedded technology.[6]

In June 2010 the new UK government announced that it would back a scheme pioneered by the Royal Borough of Windsor and Maidenhead where residents are rewarded for recycling (see box, right).

However, it's hard to see how 'chip and bin' could be made to work in inner-city areas where most of the population live in flats, many of them conversions. It's almost impossible to link households living in multiple occupancy buildings accurately with their rubbish or recycling bins (but see box, opposite, for how the Belgians get round this problem).

'Chip and bin' in Windsor and Maidenhead

In June 2009 Windsor and Maidenhead started a chip bin trial that allowed residents to earn up to £135 a year in gift vouchers for recycling. Bins were weighed and residents earned points according to how much recycling they had generated, which could be redeemed at Marks and Spencer outlets. 6,500 residents were given the chance to be part of the trial. 70 per cent of them activated the reward account. In June 2010 the scheme was extended across the borough and the new Secretary of State for Communities and Local Government, Eric Pickles, voiced support for it. The council said recycling from those involved in the trial increased by 35 per cent.[7]

Compulsion

Recycling is now compulsory in all or part of the London Boroughs of Barnet, Hackney, Hammersmith and Fulham, Harrow, Bromley and Waltham Forest, which theoretically means householders can be fined if they don't recycle. This can be done under the provisions of the Environmental Protection Act (1990). All councils introducing compulsion report a one-off rise in recycling rates. Barnet was the first to introduce compulsion in London, but its residents mostly live in single houses. As with 'Pay As You Throw', you need to be able to associate particular bins with particular households for enforcement of compulsory recycling to be possible. If that's not possible, then the risk is that fly-tipping will increase. Barnet's compulsory recycling scheme does not apply to those in households such as flats who have shared refuse facilities.

I personally prefer making it easier for people to recycle, rather than punishing them for not recycling, but this is something that community groups could lobby their council to introduce if they want.

» *If you live in an area with mostly single-occupancy houses, ask your council's recycling team to introduce the Windsor and Maidenhead 'chip and bin' scheme (see box above)*

Reduced waste and increased recycling using the Belgian Bag Method

There is a policy that might work in urban areas full of multiple occupancy buildings. In some parts of Belgium the cost of waste and recycling collection has been removed from the council tax and put into the price of special barcoded bin bags. Residents buy the bags from local shops. Bags for waste cost the most; bags for the most valuable recycling cost the least. Residents' waste and recycling is not collected unless it's in one of these specially barcoded bags. In theory it's the perfect solution because it should encourage people to produce less waste and to think more carefully about what can be recycled.

By 2007 waste was down to 150kg per inhabitant per year in Antwerp, a city of 470,000 inhabitants, which uses what I call the 'Belgian Bag Method'. As much as 61.5 per cent of Antwerp's waste was recycled.[8] That compares with 244kg per household in Camden and an overall recycling rate of 28 per cent in the same year.

The 'Belgian Bag Method' neatly gets round the biggest problem with Pay As You Throw in cities – the difficulty of associating a household with a particular bin. And it provides incentives for people to produce less waste and recycle more. It's clearly a big change, and in 2009 I was being told by those in power that it was too big a change. In 2008 DEFRA came up with some Pay As You Throw pilots, but sadly they wouldn't let me test the Belgian Bag Method because it wasn't one of their proposed solutions. Interestingly, not a single local authority came forward to test any of DEFRA's proposed methods!

» *If you live in an urban area with a high proportion of residents living in buildings in multiple occupancy, then lobby DEFRA, preferably in tandem with your council's recycling team, to be allowed to introduce the 'Belgian Bag Method' (see box above)*

Reuse – how can we persuade people to keep using stuff until it's worn out?

Once a product has been made, sold and deemed surplus to requirements by the buyer we need to find ways to reuse it. A reused product is a valuable weapon in the battle against climate change.

Going online to exchange or give away unwanted stuff

Cambridgeshire has one of the highest recycling rates in the country. As far as I can make out that is at least partly because the council has put considerable effort into education and enabling. For example, it created a free website – www.recap.co.uk – in cooperation with surrounding local authorities, which allows residents to swap or pass on items they no longer want. This was before the proliferation of community reuse websites.

Many areas now have community-led reuse sites, but what I like about Recap is that it's simple to use and, unlike Freecycle (www.uk.freecycle.org) and Freegle (www.ilovefreegle.org), it doesn't clog up your email inbox with news about the latest giveaway. Islington and Southwark were using something similar – SwapXchange – but Islington pulled out in 2009 because it was not being used as much as their local Freegle sites and I suspect Southwark may end up going the same way as it's also been hit by the rise of Freecycle/Freegle. But that's fine. Recap and SwapX-change emerged to fill a gap. Now the community has filled the gap so the council's resources are better spent elsewhere.

» *If you're a community group, and there's no reuse website in your area, then either set one up or contact the council's Recycling Department and ask if it might be able to help*

Give&Takes or Street Swap Days

Many councils hold Give&Takes or Swap Shops, which are like free jumble sales. Residents bring things they no longer want, but which they think others could use. Street Swap Days is a similar concept that apparently works well in English-speaking former British colonies such as Canada, South Africa, Australia and New Zealand. Residents put out on the pavement anything they no longer need for others to take away. At the end of the day the council removes any unwanted objects. If you want to try this in your area, the biggest objection you're likely to face from the local authority is the risk of it turning into an annual fly-tip day!

Belsize Give&Takes

Give&Takes in Belsize are now organised by the community, with the council providing advice and paying for the hire of the venue, an electrician to check electrical goods and a recycling charity to take away everything at the end. The Belsize Give&Take coordinators (accountants, management consultants, lawyers and dentists) are so ruthlessly efficient that the events go like clockwork and are hugely successful at diverting tonnes of 'junk' from landfill. And despite the 'ruthlessness' of the community organisers, almost everyone has fun. This strikes me as the perfect cooperation between a council and a community group.

A fun day out – giving away stuff and getting new stuff at the Belsize Give&Take. Photograph: Bruno Reddy

» *See if your council's Recycling Department would support you to introduce Swap Shops, Give&Takes and/or Street Swap days*

Some councils have now begun salvaging stuff that residents have discarded but which might still be useful and then selling it. Hampshire councils appear to be leading the way on this.[9] It's harder to do in urban areas where space is limited, but it should be possible for most councils to provide reuse and resale facilities.

» *Ask your council recycling team if they can create a reuse section at the local recycling centre*

Repair – how can we get residents to fix broken stuff rather than buy more?

Remade in Brixton is a community initiative for a waste-free Brixton (see box, opposite). It develops and promotes action to make better use of resources locally, focusing on efficient use, repair, reuse and recycling. Its vision is of a Brixton where there's no such thing as waste, where every resource is used to its full capacity, and every discarded item becomes raw material for something else.

Many community groups are now running workshops to help people repair or remake old clothes. Often that's in conjunction with councils. www.instructables.com is

REMADE IN BRIXTON AND LAMBETH COUNCIL

Hannah Lewis, Remade in Brixton

Remade in Brixton has developed from a community group into a fledgling social enterprise – and Lambeth Council has been our first client, inviting us to run a series of 'Remade Enterprise' workshops in secondary schools, working with students to develop their own enterprise projects based on reusing and recycling waste. We've also had input into Lambeth's successful Zero Waste Brixton funding bid – we're pleased to have got community composting facilities in there (linked to food growing on estates), and repairing and re-skilling events. We've contributed to the Brixton Sustainable Community Map (currently in development), and held an exhibition and creative reuse workshops in the council's Recycling Tent at Lambeth Country Show.

We've benefitted from the strong reputation that our parent organisation, Transition Town Brixton (TTB) had already built up with the council, and from having a couple of dedicated TTB members on the inside in the Sustainability Department. This led to us being approached by officers from three departments (Waste Management, Enterprise and Community Learning) to contribute to the above projects. Having a website that clearly sets out our agenda seems to have helped arouse interest in what we were doing.

I feel there's a way to go in figuring out how councils and community groups can work together most effectively, and a few thorny issues along the way. One is transparency: from the outside looking into a council, the structures and decision-making processes appear pretty impenetrable and it can be difficult to figure out who best to engage with and how to have an influence. Communication with the public sometimes doesn't really seem to invite contribution – some crucial consultations and open meetings have been announced at very short notice, and in general consultations restrict choices to a limited range of options. Councils haven't been designed to make the most of the collective intelligence and imagination of their community – I'd like to see a more open-source sort of council. Fortunately, quite a few individual staff and councillors are open to engagement, and some are really passionate about sustainability and community involvement.

Another area of tension is finances – as a community group we've given many hours of time and specialised knowledge for free, often in meetings where council staff were being paid for their time. As we move from a voluntary model to a social business we want to ensure we still get our voices heard where it matters, without giving too much away for nothing.

My ideal future vision for organisations like Remade in Brixton working with local councils would be one of real partnership, where we and other local third sector organisations are crucial in helping the council harness the wealth of knowledge, skills and ideas that exist in the local community. I'd like to see us all working together towards a zero waste economy, where we recognise the value of resources previously seen as 'waste' and strive to make the most effective use of all our material and human resources as part of a low-carbon future.

See http://remadeinbrixton.wordpress.com

New from old – remaking the world for our children with Remade in Brixton. Photograph: Transition Town Brixton an online community where members exchange instructions on how to produce home-made alternatives to consumer goods. I think this is an example of something where the community can do much of the work on its own but can probably get help from the council with meeting rooms, publicity and materials.

» *Try contacting the council's Recycling or Sustainability or Economic Regeneration teams for help with workshop facilities*

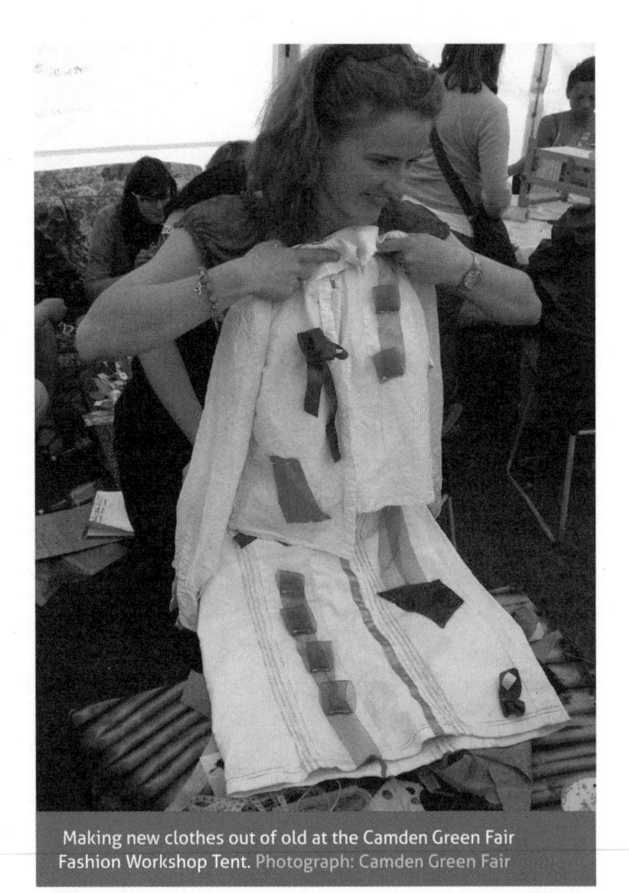

Making new clothes out of old at the Camden Green Fair Fashion Workshop Tent. Photograph: Camden Green Fair

Recycle – how can we ensure our recycling is good for the planet and that we're doing as much as possible?

The Soviet Union used to frame its Five Year Plans in terms of tonnage targets. As a result, officials had an incentive to make everything heavy. The heavier a product was, the quicker the plan was achieved. I sometimes find it hard to believe, but that is exactly what we do with recycling in the UK.

Commangled recycling – rubbish in, rubbish out

The government sets councils recycling tonnage targets, so most move to commingling – or commangling as I prefer to call it – because it's the easiest and cheapest way to increase recycling rates. Residents put all their recycling in one box and it gets crushed up in the back of a 'recycling' truck (which looks very much like a waste truck because that's exactly what it is). It then gets taken to what's euphemistically called a 'materials recycling facility' or MRF, where a huge amount of equipment and energy is used trying to separate the recycling. When the machines fail – which they do a lot of the time – manual workers are used. MRFs are getting better all the time but they're a long way from being a perfect solution.

For example, in many of them paper gets so contaminated by glass fragments and the remnants of what was in the bottles and cans that no British papermaker can use it to make recycled paper. So it gets sent to the Far East. And what happens to it there? Maybe Malaysian papermakers have lower standards or better technology. Maybe not. Either way it would make more sense not to contaminate the paper in the first place. 'Rubbish in, rubbish out' is the phrase that comes to mind.

The glass that goes to an MRF gets so crushed up and contaminated that no British bottle maker can use it to make bottles, so it's used as aggregate for roads. The carbon cost of sending glass through an MRF for it to end up as aggregate is high. It would make much more sense to keep it out of the commingled collections and give it straight to companies that make bottles or jars. And, depressingly, we create more carbon dioxide putting glass through the commingled system than we would if we simply dumped the bottles in a landfill site.

In Cambridgeshire residents are not allowed to put glass bottles in commingled collections; they have to put them in colour-separated bottle banks. These can then go straight to glass reprocessors, and in the case of the clear glass bottle bank there's actually a financial benefit to the council because there is a shortage of clear glass in the UK. The key point, though, is that in carbon terms separating glass at source is a clear win. If we all took our bottles to bottle banks instead of putting them in the commingled kerbside boxes, then we would really be helping the environment.

Recycling targets based on tonnage mean councils have a perverse incentive to collect garden waste. This largely explains why the councils with the highest recycling rates are mostly in rural areas. It is hard to see how trucking around large amounts of garden waste can make environmental sense. Composting should be done in people's own gardens. To my mind the collection of garden waste should be a chargeable service, as it is in some areas (e.g. Waverley in Surrey). The money raised could go into providing composting workshops for residents.

When I was elected in 2006 my gut feeling was that we were using more energy and emitting more greenhouse gases to recycle than we would if we did not recycle. But it took two long years before I was able to get an energy audit done that proved it.

The audit examined the collection of materials within the borough and compared 'commingled' (materials mixed together and sorted at a separate site) with 'sort at kerbside' options for collecting recyclables. The study assessed the environmental aspects, tonnage recycled and costs of the two types of collection scheme. The aim was to establish which practices were better for the environment than others and then refocus our waste and recycling strategy accordingly.[10]

The study concluded, as I suspected it would, that sending mixed recycling to a separation facility (MRF) was creating more CO_2 than it was saving. It also repeated what the Campaign for Real Recycling had been telling us for years – the separation of recycling at source creates better-quality recycling, which has more value and uses less energy to process.

The consultants who did the audit suggested some sensible solutions for improving the system, which Camden is now taking forward. More than two-thirds of our recycling is paper and cardboard. If this is separated from the other recyclables (plastic, metal, glass), then it can go direct to a paper reprocessor. They will even pay for it, which has got to be better than paying someone to ship it to the Far East, which is what Camden was doing!

Camden is also asking residents to put their glass in the on-street bottle banks wherever possible as this is by far the most environmentally sensible thing you can do with it. Add in a general food waste collection across the borough following on from a successful trial, and a collection of all kinds of plastic now that there's a recycling facility in North London that can cope with more than just plastic bottles, and I think you have a recipe for a good recycling system comprising the following five elements.

1. Food waste.

2. Paper and card.

3. Plastic, metals and glass (but only if you really can't take your glass to the on-street bottle banks because you're elderly or disabled).

4. General waste.

5. Garden waste, but only on demand and preferably for a charge.

Of course some councils have never commingled fibres (card, paper) with containers (glass, plastic, metal). Bath and North East Somerset is a good example. Derby is another. When this book went to print, Derby didn't have access to an MRF that could cope with mixed plastics so it was collecting only plastic bottles, but it did have a textiles collection as part of a five-stream collection process, as follows:

1. Food and garden waste, and cardboard (because it is organic and also reduces the quality of the paper collection).

2. Paper.

3. Plastic (only bottles), glass and metal.

4. Textiles.

5. General waste.

Councils such as Bath and Derby deserve a big round of applause for persistently doing the right thing for the planet despite being given incentives to do the wrong thing by the government.

Because of the perverse incentives given to councils to collect recycling tonnage it was only when we had the hard data from the recycling audit that we were able to change the system. And even then, because councils are evaluated on performance using tonnage indicators like this, the real challenge for councils and communities is how to reframe their waste and recycling strategies to make them more environmentally sensible, yet still hit those government-imposed targets.

If your council commingles, then I suggest you give a copy of the Camden energy audit to your ward councillor, because it's not enough to just recycle more – recycling has to actually be good for the planet

Tetra Paks

This one causes me a lot of grief. People are always telling me that you can after all recycle plastified cartons such as Tetra Paks or Pure-Paks. They know that because there are recycling bins for cartons in

Why do some councils collect plastic bottles and not other types of plastic?

If you look carefully underneath most plastic packaging you will see a small triangle with a number in it between one and seven. The numbers represent different types of plastic. Most councils can handle only types one, two and three, which are mostly plastic bottles. It is much easier to ask residents to recycle their plastic bottles than it is to explain the numbering system. Thankfully, recycling facilities are starting to be built that can deal with different types of plastic. But there are still far too many that can't be recycled. Of course an alternative would be for the government or the EU to require businesses and importers to use only materials that can actually be recycled! Would that really be so difficult?

their area. I first came across them in Godalming where my mother lives, and was stunned because for some years my partner and I had been avoiding cartons on the grounds that trying to separate the constituent parts is virtually impossible and almost certainly nonsensical.

Tetra Pak is the biggest carton company. It has an incentive to be part of the solution not the problem. That's why it put together ACE – the Alliance for Beverage Cartons and the Environment UK – to pay for carton bins. Local authorities were delighted. Carton bins count towards their recycling targets and cost them nothing.

But where do the cartons go? They're shipped to a Tetra Pak facility in Sweden. The card or fibre is removed and the plastic and metal is put through an energy-from-waste facility that runs its separation plant. I find it hard to believe that this makes sense. It might do, but nobody has yet been able to prove it to me. The one facility that used to exist in the UK – in Fife – closed down because they couldn't make it work economically even before the question was addressed of whether it actually led to fewer natural resources being used and less carbon being emitted.

My concern in all this is that we should not create situations where citizens simply feel that they can carry on doing the unsustainable thing sustainably, i.e. life can carry on much as before but with a techno-fix keeping us on the rails. It all comes back to doing recycling that is actually good for the planet rather than good for tonnage targets or good for a particular industry or business.

» *Avoid buying plastified cartons if you can, especially the ones with metal in; if you do buy them, recycle them at home by using them as plant pots or something*

Encouraging businesses to recycle

Councils have never been required to collect recycling from businesses. They only have a statutory obligation to provide a commercial waste service. As a result many businesses – especially small and medium-sized businesses – don't recycle. In 2008 Camden started offering a commercial recycling service that was cheaper than the commercial waste service. So, if a business was previously creating one tonne of waste then Camden would charge them less for creating, say, half a tonne of waste and half a tonne of recycling.

There is of course a question mark over whether councils can afford to do something like this if prices for recyclate are low, but to my mind this is mainly about quality. Good quality, which means source-separated recycling as in Derby or Bath, attracts top dollar. Low quality, as in Camden from 2006-2010, attracts bottom dollar (except that we fixed the price in advance so we got medium dollar!).

» *Push your council – the Executive Member for Recycling – to give at least your small businesses a financial incentive to recycle*

Improving the recycling experience: access, appearance, pride, competition

Recycling rates can be increased by creating more recycling centres on streets or on housing estates. Everyone should be within walking distance of at least a bottle bank. But recycling rates can also be increased by improving the recycling experience for residents in terms of the appearance of recycling facilities and incentives to use them.

The London Borough of Lewisham decorated some of its bins as cows. It then launched a 'Feed The Cows' campaign, which led to increased on-street recycling rates. Hackney later did the same thing. I believe PASSIONATELY that on-street recycling bins can and should be attractive pieces of street furniture, as they mostly are in Europe. All too often we in the UK are fobbed off with bins that are ugly to look at, revolting to use and hard to feel proud of. If our on-street recycling centres look like industrial rubbish bins (which is exactly what they are), then we shouldn't perhaps be surprised if people use them as rubbish bins and fly-tip around them.

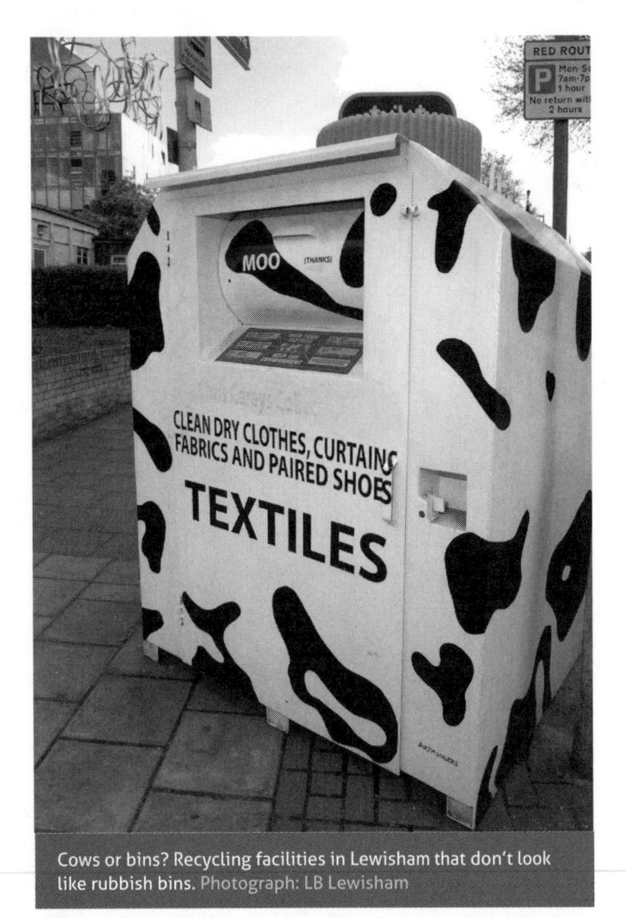

Cows or bins? Recycling facilities in Lewisham that don't look like rubbish bins. Photograph: LB Lewisham

» *Ask your Recycling Department to let schools adopt the on-street recycling centre nearest to them and decorate them as an 'art in the community' project*

Wandsworth Council put seven households through an environmental makeover, transforming them from 'wasters' to 'winners' in six weeks. The Recycle Western Riverside (RWR) 'What Not to Waste' makeover (www.westernriverside.org.uk) aimed to show how everyone could make a difference to the amount of rubbish buried in landfill sites, simply by making small changes to their lifestyles. The makeover consisted of three fortnightly challenges teaching families how to recycle, reduce and reuse the rubbish they produced.

Competitions and rewards are an interesting area of the psychology of behaviour change that all community groups could explore. But there are risks. In 2010 the new coalition government in the UK announced that people could be given financial rewards for recycling. Of course, as we've already seen, that immediately excludes everyone living in buildings in multiple occupancy (see page 150). But there's also the issue that some residents might start going through the kerbside boxes, and possibly even the on-street recycling bins, so that they can get the rewards. Competitions and rewards clearly need to be carefully thought through.

Recover – how can we maximise the extraction of energy from our food waste?

According to the public-private waste reduction group WRAP, which helps individuals, businesses and local authorities to reduce waste and recycle more, 8.3 million tonnes of food waste are produced every year in the UK.

WRAP also estimates that we throw away about one third of all the food we buy and at least half of this is food that could have been eaten.[11] That means money thrown in the bin and, if the food waste goes to landfill, which millions of tonnes do, it means the production of methane, which is more than 30 times worse than CO_2 as a greenhouse gas. It also costs councils dearly in terms of landfill charges and penalties.

WRAP calculates that 35 per cent of UK households with gardens compost either garden waste or a combination of garden and kitchen waste. There are many advantages to home composting, e.g. reduced carbon emissions from collections, less organic waste sent to landfill (although badly done home composting can lead to some methane release!), and improvement of soil quality. Those who can home-compost clearly should be encouraged to do so, and many councils and community groups offer workshops to make sure that residents get this right. However, it's important to know that councils also have a perverse incentive to pick up garden waste and cart it off to be industrially composted because that shows up as tonnage in their recycling figures.

Community groups could raise funds to supply all primary schools with wormeries. You can now buy transparent wormeries so that pupils can see the worms in action. The introduction of wormeries in schools is also likely to have a beneficial educational effect on parents, helping to increase understanding about food waste and the need to deal with it.

» *Talk to your council's Recycling Department about running wormery and composting workshops in your area, with you providing the expertise and them the publicity*

Reasons why people waste food, according to WRAP

- Buying too much – particularly being tempted by special offers, e.g. 'buy one, get one free'.
- Buying more perishable food – often as the result of trying to eat more healthily.
- Poor storage management – not eating food in date order (choosing food on impulse, often driven by 'spontaneous' and 'top up' shopping).
- Ad hoc, rather than methodical, 'spring cleaning' of stored products.
- High sensitivity to food hygiene – one in five people say they won't take a chance with food close to its 'best before' date, even if it looks fine.
- Preparing too much food in general.
- Not liking the food prepared – 22 per cent of families with children stated that not liking a meal was a cause of food waste.
- Lifestyle factors – not having the time to plan meals, or having fluid work and social patterns, particularly with young professionals.

Loving worms – Hadi from Holy Trinity Primary School in Camden proudly shows off his school's new wormery. Photograph: Polly Hancock

In October 2009 Tesco announced that it would change to a 'buy one get one free later' concept, which might mean less food waste generated in homes. But however much we use carrots and sticks with consumers and supermarkets there will still be food waste, such as that which goes past its sell-by date and which supermarkets are then not allowed to sell or to give away. Most urban households don't have gardens or outside spaces so they can't home-compost or run wormeries, and even those who do have gardens can't compost all their food waste (e.g. cooked food, dairy products, meat and fish, etc.).

So all councils should collect food waste. Here are some of the potential benefits a community group could point to if arguing for food waste collections in its area:

- Diversion of up to 30 per cent of the waste stream from landfill and therefore potentially reductions in Landfill Allowance Trading Scheme (LATS) penalties (landfill taxes).
- Reduced methane emissions from landfill.
- Increased recycling rates including paper, plastic, metal, glass, etc. because the food waste collection makes people think more about recycling in general.
- Compost creation and improvement of the local environment (parks, community gardens, allotments, private gardens).
- Use of biogas as an energy source, created when food waste decomposes (see Chapter 6, page 109).

Community composting

Some councils in London have experimented with community composting. There's a range of community-scale composting systems available. Waltham Forest, Hackney, Islington and Lambeth have all used the 'Rocket' in-vessel system to process food waste. They're called rockets because they look a bit like Stephenson's original steam engine, which helped kick off the industrial revolution that got us into the mess we're in now!

Food waste goes in one end of a Rocket composter and, after being heated to 70°C twice to sterilise the mix, it comes out of the other end as compost. In one project in East London the very effective key message to residents was that Rockets could reduce rodent levels by removing cooked food from black bags (see box, opposite).

I think this is likely to be one of the biggest issues of the coming decade. Big or small? Local or regional? Community or council? Do we truck waste and recycling miles away or do we try to deal with it on or near where it is produced? How do we define the community benefits of small-scale facilities when in terms of kilos processed for pounds spent they are more expensive? That is the challenge for community groups in the face of a long-term squeeze on council budgets.

Yer-uck! Camden Council councillor Chris Naylor examines the inner workings of a Rocket composter.

I believe that there are many community benefits to dealing with something like food waste locally in terms of enhanced resilience, increased community pride, local job creation, creating a virtuous cycle of using food as a resource, raised environmental consciousness and added incentives to grow food, but it will be a struggle trying to find ways to argue against lowest cost per tonne.

» *Create a community plan for dealing with your food waste and returning it to the soil before the council does it for you and goes for a 'big is beautiful' scheme*

Rockets reduce rodents

In May 2007 I went to visit an estate-based composting project in the London borough of Hackney. Actually it wasn't really about composting – or at least that's not the main attraction for residents. It's about rat reduction! Compost is simply a fabulous by-product.

The Nightingale Estate in Hackney is right next to a railway line and as a result used to be infested with rats. However, since they installed their 'Rockets', which turn food waste into compost, they have almost completely eradicated their rat problem. Food waste is collected in rat-proof caddies and so is no longer left in plastic rubbish bags outside flats or in bin stores. Using the vermin eradication message led to participation rates of up to 80 per cent after a door-to-door education campaign. This dropped to 50 per cent six months later, when a further door-knocking campaign brought the rate up again.

The head of the Tenants Association told me the rats had gone and that the compost had been used to help regenerate the estate. Since food is about a third of all household waste the environmental benefits of recycling it locally are huge. However, the costs of installing a Rocket are higher than a standard food collection and sadly Hackney Council failed to find ways to justify an expansion of the incredibly successful Nightingale Estate experiment.

WHY I LOVE MY ROCKET

Rachel Zatz, resident of Maiden Lane Housing Estate, London Borough of Camden

I'm delighted with the Rocket Composter we have on our estate because it means I produce almost no rubbish now – just one bag of waste every three weeks. Being able to get rid of cooked waste has made a huge difference. My bin bag can now be left in for longer. People on the estate are also telling me it's made them realise just how much food waste they were creating.

The Rocket was the idea of councillors but there had already been an estate needs survey, which resulted in 74 per cent of respondents saying 'yes' to food waste collections, so we knew there was underlying support for the concept. That said, many residents will probably feel that the Rocket was imposed on the estate because it wasn't mentioned in the survey, which is a shame in terms of the sense of cooperation between the council and residents.

The Rocket could be a really positive liaison force between residents and the council. It gives residents pride. But can't be about blackmail, i.e. we'll give you a Rocket if we can knock down half the estate! [Maiden Lane was going through a controversial estate regeneration scheme at the time the Rocket was installed.]

Then there was the smell problem. I actually thought it was a good thing because: a) it raised the profile of the Rocket; and b) the council did something about it. [Smells were not a problem once some modifications were made.] The Rocket is a really positive thing for the estate. I'd like to see us get up to 80 per cent participation.

They're supposed to employ one of the residents to work on the Rocket but nobody has taken the job yet. The question of what happens after the two-year trial is still up in the air. It really should have been locked down in advance. Hopefully there will be enough councillors prepared to fight for it in two years but it might be a big fight. A one-size-fits-all policy is just not right. There are non-financial community benefits, which I think the council should pay or, if this community wants to make a go of it, then it should be supported to make a social enterprise.

Chapter 11

TRANSPORT

Questions answered in this chapter:

- Transport in transition – is it about technology or relocalisation?
- How can we create safer, healthier, happier and lower-carbon neighbourhoods?
- How can we persuade people to get out of their cars?
- What transport policies could be reframed in terms of carbon emissions?
- How can we get more people cycling?
- What can we do to decarbonise freight?
- Can communities and councils do anything about aviation emissions?

I have a vision of my neighbourhood in London, which is all about reclaiming the streets from the motor car. The world's 800 million cars go unused a staggering 88 per cent of the time. They clutter up driveways, car parks and streets. Just think of the trees, the green space, the properly separated cycle lanes and the widened pavements we could put in if we got rid of a significant proportion of our parking spaces. There might even be a rainwater-fed stream as in the South German village of Frickingen (see photo).

The only cars left in my ideal neighbourhood would be car-club vehicles and they would be non-polluting, electric or biogas cars restricted to a maximum speed of 20mph in residential and shopping districts. Trucks would be banned altogether. Delivery vehicles would be smaller and they wouldn't be diesel-powered. All residential roads (apart from cul-de-sacs) could be cut in the middle, because nobody needs to drive through them, so the middle section could be turned over to food growing. Shopping streets would basically

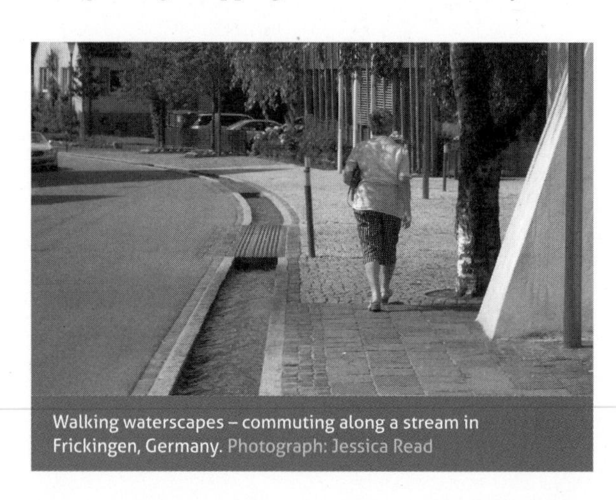

Walking waterscapes – commuting along a stream in Frickingen, Germany. Photograph: Jessica Read

become large pavements, with the road raised to the height of the pavement and made of a slightly rough surface such as brick to signify to drivers that they are no longer in motor-car territory.

Pupils would be able to get to school safely on foot or by bicycle. The air quality would be better and children would become less prone to asthma attacks. We would become fitter and experience better health in retirement. Village, town and city centres would become more pleasant for residents and more attractive to visitors.

Councils have the power to remodel streets like this, but they need the buy-in of residents. Community groups can play a critical role in terms of creating buy-in by spearheading visioning exercises for their area. In Camden in 2008 I brought together a group representing residents, parents and cyclists and we created a vision of how Fitzjohn's Avenue, the street most affected by the school run, could be transformed using echelon parking, separated cycle lanes, car-club spaces, enforced speed limits, an electric train on wheels (such as you see in tourist hotspots like Avignon in France) and school buses. We didn't get everything we asked for, but I'm sure that we ended up with a 20mph limit, average speed cameras, car-club spaces and school buses because so many community groups were involved in the visioning process.

Going electric – an electric train for tourists in Avignon.

Do It Yourself Streets

DIY Streets is a Sustrans project to help residents to redesign their own streets affordably, putting people at their heart, and making them safer and more attractive places to live.[1] The project works with local communities to help residents develop low-cost capital solutions to making their streets safer and more attractive, aiming to find simple interventions and materials that can be both effective and durable.

The approach is initially being piloted in 11 communities, with the intention of becoming replicable on a national scale in the near future, delivering the benefits of people-friendly streets at a fraction of the typical cost of a home zone. A pocket guide and additional design information sheets are also available. They offer advice and tips to those wishing to apply the DIY Streets approach to their own street.

See www.sustrans.org.uk/what-we-do/liveable-neighbourhoods/diy-streets

Community groups can do a lot by deciding for themselves what they want and then lobbying for it, e.g. for a new bus route, or just getting on with it, e.g. a local car-share scheme. Often a council can provide enabling power if you can provide the ideas and energy. That's what happened with Camden's school bus scheme – parents organised the buses and the council facilitated the planning changes and parking issues.

» *Organise a campaign to redesign your neighbourhood in transport terms and get your ward councillors on board, then take it to your council's Transport Policy Department and Sustrans*

TRANSPORT IN TRANSITION – IS IT ABOUT TECHNOLOGY OR RELOCALISATION?

Peter Lipman, Policy Director, Sustrans and Chair, Transition Network

How, and how far, will we travel if we make the changes we need to in order to thrive in a carbon-constrained society? For a range of interlocking reasons, I believe we will be happier, healthier and more resilient if we radically change from our current patterns to ones that fit into a relocalised world. In that world we will travel far less far and fast, overwhelmingly walking, cycling and using public transport.

Underlying our choices about how to travel is why we choose to travel at all. Although some of us occasionally travel for its own sake, the majority of the trips we make are to access something we need (schools, shops, workplaces, parks, etc). A quick trawl of travel data from the National Travel Survey[2] reveals that we're going to the same places we used to 50 years ago, but we're travelling further to get there. A case in point is the school journey; as the government extends our choice as to which school is attended, we end up with less choice on how to get there – the average school journey has increased from 2.9 to 3.3 miles in the last few years.

This means that if we want to move to a world in which sustainable modes of transport dominate, we have to ensure that the locations we all need to access in order to prosper and thrive are within reach by foot, bike or public transport. At the same time we have to think very hard about the kinds of physical infrastructure we create, as the environment we create impacts enormously on the choices we then make or feel able to make.

If, in addition to having a long way to go to reach our destination, we encounter a hostile environment when we step out of our front doors, we'll tend to react defensively, often retreating into what seems to be a safe refuge of a car. On the other hand if we emerge into a space which welcomes people generally (not just travelling but also for example socialising and playing) then we'll tend to react expansively, feeling able to walk or cycle. But of

How far will we go? Pete Lipman, Policy Director of Sustrans, addressing the 2010 Transition Network Conference. Photograph: Mike Grenville

course it won't help if that welcoming environment comes to an abrupt stop at the end of our street – so it needs to continue all the way to our destination.

Dramatically cutting emissions while still continuing to travel further and faster demands a technological fix. When the International Energy Agency ('Energy Technologies for a Sustainable Future: transport') reviewed this subject, it concluded that: "There are only three basic approaches to achieving a transport system with very low emissions of greenhouse gases and low reliance on fossil fuels: a hydrogen fuel-cell system; a purely electric vehicle system; or relying on liquid fuels derived from biomass."

Any system based on one or a mix of these measures would take time to implement, when the need for very significant reductions is urgent. In addition, technical solutions may create new, even worse problems. For example, demand for agrofuels as the substitute for oil-based fuels plummeted as it became clear how they compete with food production. In 2006, the first year in which the US turned more of its corn into ethanol than it exported, tortilla prices in Mexico tripled, and food riots followed. Similarly, the switch to agrofuels led to a rush to establish palm plantations for purportedly 'climate friendly' palm oil. The result was very significant rainforest destruction, and all that implies for biodiversity, and enormously increased (up to 15 times) overall climate change emissions from the clearing and burning of the forests.

Visions of enormous fleets of 'clean' electric or hydrogen or hydrogen-fuel-cell-powered cars raise a range of questions, such as just how much extra energy will be needed to construct the necessary new infrastructure? A truly clean car would require all stages of its life to have been powered by renewable energy, from the mining of raw materials, through its manufacture, shipping, sale and disposal, as well as for each and every electric charge used to power it. In an energy-constrained world would we really choose to power cars over hospitals and homes?

Around the world, on average we make about 1,000 trips (e.g. from home to work – that's one trip), by any mode of transport including by foot, per person per year. Travel behaviour research from across Europe, the United States and Australia consistently shows that 10 per cent of people's car trips are shorter than 1km, 30 per cent are shorter than 3km and 50 per cent are shorter than 5km. This large number of small trips means that, even before relocalisation really starts to take hold, we have the potential to immediately intervene to support more cycling and walking trips – much more quickly than for any technological development and at a fraction of the cost. In fact, even under current conditions about half of the car trips we make could be switched immediately to sustainable modes.

If, however, we are to consider technology-based solutions, we need to learn to ask ourselves far harder questions than we've managed so far, including:

- What is the full energy cost of this transport intervention, including all of the embodied energy in the infrastructure needed as well as that from running the system?
- How long will it take to implement; could carbon reductions be achieved any faster with a different intervention?
- Do we, as a society, have enough energy overall to carry through our decision?
- Similarly, do we have enough money to carry out our decision?

Applying such an analysis might result in a transformed UK, in which we've maximised the use of our existing infrastructure and:

- nearly all urban trips are on foot, by bike or by biogas-fuelled public transport
- rural trips that can't be done on foot or by bike are mainly by community-owned demand-responsive vehicles, again biogas-fuelled
- longer trips are mainly on an electrified rail network or by coach.

We don't know whether humanity's climate change emissions so far have caused a soluble problem or an insoluble predicament. We may already have pumped sufficient carbon into the atmosphere to have triggered feedback loops which will lead us well beyond a 2°C temperature increase – and even a 2°C increase could turn out to be far more dangerous than mainstream climate literature predicts.

Accordingly, applying the precautionary principle and minimising unquantified risks, we should be seeking urgently to move to zero-carbon travel, which could happen fastest in a relocalised world. Does this mean forgoing the attempt to somehow find a technological fix through which we could continue travelling as far, and as fast, as we want? While that might seem hard to contemplate, we probably don't have the choice – and in addition, we'd also address the fact that a result of current travel patterns is that we're rapidly getting less healthy and less, rather than more, happy.

How can we create safer, healthier, happier and lower-carbon neighbourhoods?

Shared space

Where volumes of traffic are reasonably low, raising the road up to pavement level and removing road markings can reduce traffic speeds and demonstrably increase pedestrian and cyclist safety. The principle is one of shared ownership – if shared space is visibly different from normal road space, then pedestrians and cyclists feel they have just as much right to use it as drivers. Of course some councils have gone further and introduced car-free areas at certain times, e.g. Saturdays or market days. This sort of temporary traffic ban was successfully implemented by the London Borough of Westminster in Oxford Street. It changes the dynamic utterly. On 5 December 2009 two streams of people took over the streets of central London – the shoppers on Oxford Street and the climate protesters marching to Parliament Square. It reminded me of my time living in Paris where, every Friday evening, between 3,000 and 10,000 roller-bladers took over the streets for a police-escorted 30km tour of the French capital.

Filtered permeability

Filtered permeability is the concept that networks for walking and cycling should be more permeable than the road network for motor vehicles. This, it is argued, will encourage walking and cycling by giving people a more attractive environment free from traffic and a time and convenience advantage over car driving. Evidence for this view comes from European cities such as Freiburg and Groningen, which have achieved high levels of walking and cycling by following similar principles, sometimes described as 'a coarse grain for cars and a fine grain for cyclists and pedestrians'.

Filtered permeability requires cyclists, pedestrians (and sometimes public transport) to be separated from private motor vehicles in some places, although it can be combined with shared-space solutions elsewhere in the same town or city.

The principle of filtered permeability was endorsed for the first time in British government guidance for the eco-towns programme in 2008, and later that year by an alliance of 70 organisations concerned with public health, planning and transport in their policy declaration 'Take Action on Active Travel'.[3]

A parallel debate has been taking place in North America, where some commentators have proposed an urban form that follows the principles of filtered permeability – the Fused Grid – to address perceived shortcomings of both the 'traditional' grid and more recent suburban street layouts.

Reducing speeds

Lowering speeds makes streets safer – no question. According to an oft-quoted 1979 survey, for each 1mph

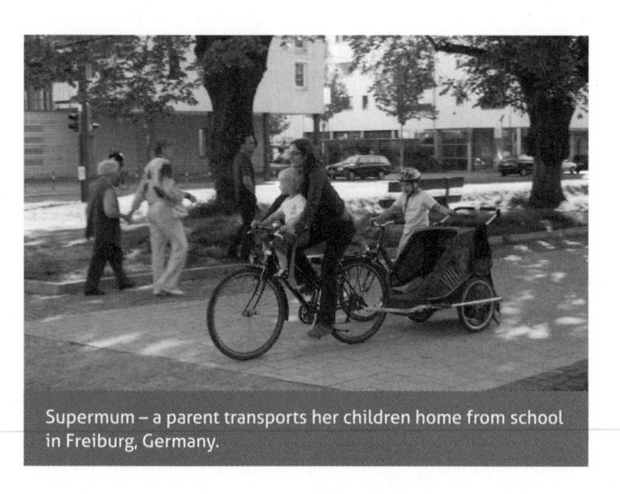

Supermum – a parent transports her children home from school in Freiburg, Germany.

US streets aren't made for walking

The United States almost certainly has the biggest transport problem of any industrialised nation because so much of it was built for the car. The peak oil film *The End of Suburbia* paints a bleak picture of the collapse of the US dream when oil becomes prohibitively expensive and suburban life becomes impossible because there are almost no public transport links.

The issue is belatedly being addressed by the Obama administration. In a dramatic change from previous policy, US Transportation Secretary Ray LaHood proposed in January 2010 that new funding guidelines for major transit projects be based on liveability issues such as economic development opportunities and environmental benefits, in addition to cost and time saved, which had been the primary criteria.

"Our new policy for selecting major transit projects will work to promote liveability rather than hinder it," said Secretary LaHood. "We want to base our decisions on how much transit helps the environment, how much it improves development opportunities and how it makes our communities better places to live."[4]

The change will apply to how the Federal Transit Administration evaluates major transit projects going forward. In making funding decisions, the FTA will now evaluate the environmental, community and economic development benefits provided by transit projects, as well as the congestion relief benefits from such projects.

reduction in average speed, accident frequency is reduced by 5 per cent.[5] Also, if a pedestrian is hit:

- At 20mph there is about a 1 in 40 (2.5 per cent) chance of being killed or 97 per cent chance of survival.
- At 30mph there is about a 1 in 5 (20 per cent) chance of being killed or 80 per cent chance of survival.
- At 35mph there is a 50/50 chance of being killed.
- At 40mph there is about a 9 in 10 (90 per cent) chance of being killed or 10 per cent chance of survival.

In 1999 local authorities were given the powers to introduce a 20mph speed limit without requiring the consent of the Secretary of State. In 2007 Portsmouth brought in a blanket 20mph on residential roads across the city, the first city in the UK to do so. Since then Oxford, Norwich and Aberdeen have all decided to go down the same 20mph road. Strong residents' campaigns were key in all cases.

Research done on 20mph zones in 1996 by the Transport Research Laboratory[6] suggests:

- Average speeds within zones reduce by 9mph.
- Accident frequency reduces by 60 per cent.
- There is an overall reduction in child accidents of up to 67 per cent.
- There is an overall reduction in cycle accidents of up to 27 per cent.
- Traffic flow within zones is reduced by up to 27 per cent.

There's some evidence that simply putting up 20mph signs can reduce speeds, but a 20mph speed limit can be really effective only if it's introduced in conjunction with traffic-calming measures. Unfortunately national police policies are against using resources to monitor and enforce 20mph speed limits, so that means it's up to councils and residents to do the work. Most enforcement is about cheap humps, which lead many drivers to slow down and then speed up, thereby creating more CO_2. Furthermore, car engines are currently not designed to run efficiently at 20mph – although they could be and certainly would be if they were electric!

» *Organise a petition of residents to reduce the speed limit in your area to 20mph and then take it to the council*

Average speed cameras

Average speed cameras have been used to good effect on motorways to slow down traffic. In 2009 the Department of Transport gave permission for four average speed camera pilots in urban areas, so this should be a possibility for all at some point in the future.

Installing average speed cameras means more street furniture and photographic databases, which create civil liberties issues, but I believe the benefits in terms of reduced speeds and increased safety outweigh these concerns. But the main obstacle to roll-out is likely to be the expense (and the new coalition government!) – they cost up to £500,000 for a system, depending on how many cameras are installed. That unfortunately means that on residential roads humps and chicanes are likely to remain the single most cost-effective way of reducing speeds – until we can persuade residents to create food-growing sites in their roads!

Camden Council has been testing a 20mph zone in Camden High Street using traffic lights. Vehicle speeds are controlled by coordinating the timings of the various sets of signals to allow traffic to progress at no more than 20mph. Those exceeding this limit get no 'green wave' and therefore would have to stop at existing signals.

This pilot is encouraging because it uses existing signals to do something that has additional benefit. In Switzerland they use stand-alone traffic lights, located away from junctions, which turn red if a motorist exceeds the speed limit. The lights are radar-controlled. If the driver jumps the red light, he or she is photographed and fined.

» *If car speeds are a problem in your area, then start a campaign to persuade your council's Transport Policy Department to install average speed cameras or stand-alone radar-controlled traffic lights*

How can we persuade people to get out of their cars?

The main step we took to reduce traffic in Camden was taken before climate change was a concern. It was the introduction of Controlled Parking Zones, which meant residents needed a permit to park in Camden's streets. At a single stroke this stopped people from outside London commuting in to Camden in their cars. The rest, frankly, has been peanuts. Car-free housing, emissions-based parking charges, car-club expansion – all good, but we haven't done enough to make much of a dent in our transport emissions. We need to bite the bit and take vehicles off our streets.

In rural areas people are going to struggle without cars. I'm not a big believer in the hydrogen revolution, so it looks to me like the solution will have to be a combination of more shared vehicles (see box on page 171 for Devon's excellent car-share scheme) fuelled by biogas or electricity from renewable sources and a return to the horse and cart!

In urban areas most people should be able to make do with a combination of more walking, cycling, public transport and car clubs. The disabled and the frail are really the only exceptions. But people think they need a car. It's become a human right. No matter that you spend more and more time stuck in traffic, that there is increasingly nowhere to park when you arrive, that you inevitably get parking tickets and that road rage is on the rise.

Car clubs

The UK's largest low-carbon community, BedZED in Surrey, secured its biggest carbon reductions not from energy-efficient walls or a wood-fired combined heat and power system, but from its car club. Not only were residents sharing cars, they were using them less. On average a car club car replaces ten privately owned cars. For those who doesn't use their car every day, they make financial sense. You pay a small annual fee to join and then per hour used. You book over the phone or on the internet. A swipe card and a PIN code gives you access to the car and allows you to drive it.

The car club looks after insurance, servicing, MOT, cleaning, etc. If there's no car available exactly when you want it, or if you're in a terrible hurry, then most people would still be better off financially by using the occasional taxi.

It sounds good, but joining a car club isn't an obvious step for most people. Persuading them to submit to the possibility that they won't be able to use a car exactly when they want to is hard work. It's right up

Smarter Travel Sutton

In 2006 Sutton Council's Sustainable Town Centre Pilot Project, known as 'Smarter Travel Sutton' began seeking to achieve a modal shift in transport use in the borough and hence a reduction in greenhouse gas emissions from road traffic, by:

- encouraging non-car modes, including walking, cycling and public transport
- encouraging use of closer destinations and facilities
- changing the time of travel, moving from peak to off-peak where possible
- reducing the need to travel for some journeys (e.g. internet shopping and teleworking)
- encouraging better use of transport resources (e.g. vehicles, fuel, etc.) where sustainable modes may not be viable for certain journeys.

The Smarter Travel Sutton project is focused on promoting personalised travel planning, school travel planning and workplace travel planning throughout the borough. That means a lot of resources put into face-to-face work. The initial results, released in February 2010, were impressive.

There was a 75 per cent increase in cycling in Sutton, a 16 per cent increase in the number of people travelling by bus, and a 6 per cent reduction in the proportion of residents' journeys made by car.[7]

All of Sutton's 68 schools had a travel plan in place by March 2008, two years ahead of government targets. More than 16,000 employees in Sutton work for organisations that have a travel plan. Every household in Sutton has been offered free, personalised travel advice and information to help them make informed choices about the way they choose to travel. In addition, 27,000 people attended Smarter Travel Sutton events throughout the course of the final year, 10,000 primary school children regularly participated in the WoW (Walk Once a Week) initiative and over 1,000 Sutton residents have had a first appointment with an Active Steps advisor in an initiative that aims to improve health by encouraging people to change their travel habits.

See www.smartertravelsutton.org

there with trying to persuade people to fly less often or to eat less meat.

Camden used to 'sell' car-club spaces to residents via a consultation that said "We're thinking of plonking a car-club space in your road. It will mean the loss of a resident's parking space. Yes or no?" No prizes for guessing which way most residents voted. I eventually managed to get the car-club consultation documents changed so that they stressed the benefits of car-club membership. I call it mindset marketing!

In theory car clubs can work perfectly well in rural areas as well – every village could have a car-club space. In practice I suspect the established car clubs would struggle to make an internal business case if the usage volumes were low. You might need to set your own car club up. Contact the council's transport team and see if they can help. They might be able to rent you a vehicle.

Commonwheels is a not-for-profit car-club company that would be worth contacting if you want to try setting one up in your area. See www.commonwheels.org.uk.

In urban areas nobody should be more than a few minutes' walk from a car-club space. Community groups can push for car-club spaces by starting petitions of residents. They can also help with 'mindset marketing' by circulating information about the benefits. Meanwhile, councils can push the car-club companies to make their offers more flexible, better value and more convenient.

I know parents who would be prepared to give up their family cars if some of the car-club vehicles were big enough. I know lower-income families that would give up their old bangers if car-club cars were cheaper. Car clubs are not yet well adapted to all, but I think they are the future of most driving.

This is the place to go if you want to find car clubs operating in your area: www.carplus.org.uk/car-clubs/find-a-car-club-car.

» *Raise awareness in your neighbourhood about the benefits of car clubs, start a petition and lobby your ward councillors for a manifesto commitment to the idea that nobody should be more than a few minutes' walk from a car-club space*

Car-free housing

New developments are sometimes designated car-free by councils if they are easily accessible by public transport; near a range of amenities, including shops and leisure activities; and within a highly stressed controlled parking zone.

For those living in car-free housing schemes no on-street parking is permitted, except for disabled drivers. Car-free housing developments are secured through planning agreements between the council and developers through Section 106 of the Town and Country Planning Act 1990, known as S106 agreements. When you've become an expert on planning, you can push for this at Development Control meetings.

It's possible for houses or housing blocks to elect to go car-free voluntarily, by asking the council to change any Traffic Management Orders associated with the building, but the building owners would have to bear any costs incurred by the council and it could potentially devalue the property so it's unlikely that anyone would do it.

» *Make sure your council's Planning and Transport Policy Departments are working together to create car-free housing wherever possible*

» *See whether you can set up a car-share scheme and ask the councils in your area to back you to give it credibility*

Car exclusion zones and road pricing

Children are healthier and happier when they walk or cycle to school and their parents generate fewer carbon emissions.[8] That's why, in 2007, the European Environmental Policy Institute suggested introducing car exclusion zones around schools to encourage walking and cycling.[9] When I looked at this in Camden I found that some schools were already operating car exclusion zones informally. I tried to persuade the council to introduce this throughout the borough but sadly nobody thought it was a realistic proposition in central London!

» *Community groups should discuss car exclusion zones with schools before presenting a united front in discussions with the council*

Berlin has a system of traffic-light badges for cars – red for the most polluting and green for the least. Some areas only allow green-badged cars in. There are discussions in a number of European countries about using GPS technology to charge for access to particular areas, e.g. around schools or heavily congested junctions, using GPS technology and charging more for cars that pollute more. The British government tried to start a discussion about introducing such a system in 2005 but was forced to abandon it in the face of public protest led by the Drivers' Alliance.

Devon's car-share scheme

Devon County Council have worked in partnership with Plymouth City Council and Torbay Council to set up www.carsharedevon.com. This website aims to maximise people's travel options whilst also reducing the number of cars on the roads, cutting pollution, saving money and protecting the environment.

The benefits given by the council for car sharing are as follows:

- Saves you money – travelling with others enables you to reduce your transport costs by up to £1,000 a year.
- Reduces the number of cars on the roads – resulting in less congestion, less pollution and fewer parking problems.
- Provides a real solution to the transport problems of rural areas.
- Gives employees and employers more transport options.
- Reduces the need for a private car.

See overleaf for a case study in the car-share scheme.

The power to introduce road-user charging was given to the Mayor of London in the Greater London Authority Act 1999. The first Mayor, Ken Livingstone, introduced a congestion charge in 2003 in an attempt to discourage people from driving into the centre of London. It led to a 15 per cent reduction in traffic when it was introduced but volumes have since returned to pre-charge levels. Mayor Livingstone wanted to turn it into an emissions-based charge but his successor Boris Johnson vetoed the move.

Other UK councils need approval from the Secretary of State for Transport to introduce road pricing or congestion-charge schemes. Only Durham has done

Julia – a case study in car sharing by CarShareDevon

When Julia moved from Surrey to a new job in Exeter, for one reason or another she ended up living 50 miles away in Launceston. That meant a round trip of over 100 miles every day. Julia contacted CarShareDevon to see if they could find her either a lift or people to share her car. They absolutely could. And in the process Julia cut her travel costs, reduced her stress levels and made new friends, which helped her to settle in to the area. With all four seats usually taken, and cars and driving rotated, the bills have been cut drastically. "I used to fill up twice a week," said Julia. "Now I fill up about once every two weeks." Julia and her car sharers have calculated that they are each individually making an annual saving of just under £3,000 on the costs of fuel and vehicle maintenance. Julia and her car-share colleagues work out a weekly driver roster and have located mutually beneficial meeting spots en route with approved secure areas to leave their cars, if they're not left at each other's homes.

"I thought the distance and remoteness of where I live would make it hard to find people to share with, so I was amazed when I registered with carsharedevon.com to find so many people making the same journey as me and at the same times.

"None of us knew each other before and the website gave me confidence to safely communicate with potential sharers. It is good having people to talk to from different jobs and backgrounds and being new to the area I have been given lots of good advice. We have even found we share similar interests such as animal welfare and gardening as hobbies. Sharing with others has enabled me to stick closer to my contracted hours and separate work from home life more easily, which in turn creates improved health and productivity. As car sharing becomes more popular, organisations should be ensuring measures are put in place to provide backup to car sharers so they can get home in an emergency, as this is often a concern which prevents people considering car sharing as a travel option.

"When my car wouldn't start one morning, one of the others came to pick me up and continued to go out of their way to collect me until my car was repaired. Without car sharing I wouldn't have been able to get to work until my car was fixed. My advice is not to dismiss car sharing just because you live remotely or commute over a long distance. You'll be surprised at how many others make the same journey".

CarShareDevon 2010

so. Proposals for road pricing in Cambridge, Edinburgh, Manchester and the Midlands have been overwhelmingly rejected by voters.[10]

Old cars

Old cars are a source of significant emissions, so incentives introduced to persuade residents to scrap old cars should be a good thing. Not always! In 2009 the government announced a £2,000 subsidy to anyone buying a new car as a way to save jobs in the car industry. A huge opportunity was missed to kickstart a move to more fuel-efficient cars or indeed out of cars altogether.

» *Ask your ward councillors to sponsor a policy of giving those who scrap their old car, and who commit to not buying a new one, a voucher that gives them membership of the local car club*

Workplace parking levy

Under the Transport Act 2000 councils have the right to apply a charge to any business that provides parking for its employees – a workplace levy (WPL). It cannot be applied in areas that have a congestion charge and it cannot be applied to supermarket car parks (sadly!).

Nottingham will be the first authority in the country to introduce a WPL. The authorising legislation leaves it to the local authority to decide precisely how such a scheme will operate, so a lot of work went into designing a legally sound authorisation process for the Nottingham scheme. Widespread consultation was undertaken in 2007 and as part of this exercise an innovative one-week 'public examination' of the proposals was included to provide enhanced public participation and allow for independent scrutiny.

In July 2009 the Secretary of State for Transport confirmed the legal order, opening the way for the Nottingham WPL scheme to proceed. It will start in October 2011, although charging for licences will not begin until April 2012. That will be 12 years after the concept of a WPL was introduced into law and more than five years after Nottingham started working on the plan! Still, the heavy lifting has now been done, so community groups could lobby for this as a way to fund green transport measures.

Ask your council's Executive or Cabinet Member for Transport to consider a workplace parking levy

Darlington Sustainable Transport Town

This was a five-year project (2004-09) funded by central government to try to reduce transport levels by talking through options face-to-face with people. All households were given a chance to participate. In addition, 30 schools and 23 businesses and organisations completed travel plans. More than 1,000 bike stands were also installed in 20 schools.

Traffic levels fell by 9 per cent in the first two years. Walking saw the biggest gains in absolute terms, with an additional 38 trips per person per year being made on foot, a relative increase of 15 per cent. Cycling gained an average of 18 trips per person per year, showing a relative increase of 120 per cent.

The total cost was £3.3m, which was spent on interviews, doorstep travel advisors and cycling infrastructure. The scheme is estimated to save 7,000 tonnes of CO_2 per year compared with a 2004 baseline.[11]

Buses

As long as buses are reasonably full and/or are running on a genuinely green fuel, then moving people from cars into buses makes a lot of sense. In some parts of the country bus services are tightly regulated, as in London or Northern Ireland. In others there are as many providers as the market will bear. In the countryside bus services are usually subsidised for social reasons because private operators find it hard to make a profit. In theory, outside London (Transport for London) and Northern Ireland (Translink), anyone can set up a bus service as long as they satisfy the local Passenger Transport Executive (larger conurbations) or the local Transport Commissioner. But even in London there are exceptions to this rule, as in Camden, where parents and the council worked together on setting up a school bus service to alleviate the problem of the school run.

If there's a potential bus route in your area that you think would work, then ask the Transport Planning Department of the local transport authority (probably the county council or borough council) to help you set one up

Transport for London has commissioned ten buses running on hydrogen fuel cells at a cost of more than £1m each. If you can make the hydrogen without using fossil fuels, then the output of a hydrogen-powered vehicle is just water. But for the moment most hydrogen is made from natural gas. In the case of London's hydrogen buses they're made in Northern Ireland, then sent to California to have the engine fitted, then sent to London.[12] This feels like expensive, techno-fix escapism. It would make much more sense to run buses on genuinely green transport fuels such as electricity made from renewables or biogas made from food waste (or crop waste, animal slurries and sewage).

The Brighton and Hove bus partnership

Brighton Council and the local bus operator have worked together since 2000 to increase bus usage and reduce traffic. The deal isn't a formal commercial one – there's some agreement on which buses should run where and when, but basically the council concentrates on investing in facilities such as bus stops and the operator invests in buses.

- Council commitment: bus priority of road space, improved passenger waiting areas, real-time information displays, traffic regulation enforcement, provision of park & ride.
- Bus company commitment: improved service frequencies, value-for-money fares and tickets, investment in new buses, enthusiastic staff, effective sales messaging.

Traffic levels have been reduced by 10 per cent since 2000 and there was a 33.5 per cent increase in bus use between 2000/01 and 2008/09. And between 2000/01 and 2005/06 passenger satisfaction increased from 56 per cent to 80 per cent.[13]

The Adelaide Solar Bus

Adelaide City Council introduced the world's first solar electric bus – the Tindo, which is the Kaurna Aboriginal word for sun. The Tindo's batteries are recharged from the large array of solar panels on the roof of the Adelaide Central Bus Station. This is estimated to save around 30 tonnes of CO_2 equivalents per year from avoided diesel emissions, and, according to the council, has raised considerable community excitement about sustainable transport.[14]

» *Organise a workshop in your area on electric and biogas buses and invite representatives of your local bus companies and the council's Transport Policy Department*

What transport policies could be reframed in terms of carbon emissions?

The London Borough of Richmond was the first local authority in the country to introduce differential parking charges based on the size of vehicle engines. The launch attracted huge media interest, at least partly because virtually everyone in Richmond seems to have a large car! The charge for a resident's parking permit for the largest engines is £300 per annum or £450 for a second or subsequent car. Hopefully that's high enough to nudge behaviour in the right direction without provoking too big a backlash by drivers. Just as important will be the public's need to see the proceeds spent on green transport measures. The answer may well come at the ballot box.

Richmond has since introduced CO_2-related charging in all its car parks, using new smart meters and the RichmondCard, a standard-sized plastic card with chip that stores vehicle registration and CO_2 tariff band, as well as cash, which can be used to pay for parking. I would like to see Penalty Charge Notices varied according to engine size. For the moment London Councils, the body that represents the interests of London boroughs to government, has rejected this. But there's nothing to stop councils outside London trying it. And there's nothing to stop community groups campaigning for all parking charges to be reframed in terms of carbon emissions.

» *Lobby your council's Parking Department to reframe its parking policy based on emissions*

Electric cars

Many councils, particularly in urban areas, have brought in incentives to make electric-car ownership more attractive. In some areas electric-car owners can obtain a free or reduced parking permit if they can prove their domestic electricity supply comes from a renewable energy provider.

Camden did a lifecycle analysis of vehicle fuels that included electric cars. It concluded that unless renewable energy was used to power an electric car, it would create more carbon over its lifetime than a small diesel or petrol car, because of the energy used to make the battery and because of the carbon intensity of grid-supplied electricity – the fact that most of our power comes from gas, coal and oil-fired power stations. This last point is the reason why electric cars are not going to solve our carbon problems any time soon. There's simply not enough electricity from renewable sources in the UK to cope with a switch from the internal combustion engine to electric, and there's unlikely to be for decades.

Some councils in London have installed on-street charging points, but not enough to really make a difference. The problem is that there's not enough on-street parking to tie up one space for the time that a resident would need to charge their electric car. For the moment, battery charging takes too long. As a result the only people who can really have electric cars in urban areas with serious pressure on parking are those who have off-street parking.

This is likely to change as battery technology gets better. I personally think the solution will be portable

Could Zermatt shine an electric light on the future of urban villages?

Taking a winter break from eco-evangelism in Zermatt in Switzerland, I was struck by how a beautiful mountain village in the shadow of the Matterhorn might hold the electric car keys to the future of urban areas. For Zermatt is a village without cars. Or, to be more precise, there are no fossil-fuel-powered cars. The only motorised vehicles in Zermatt are electric and they can be used only by businesses and essential services. Even the police ride around in electric vans.

The first-ever cars, back in the nineteenth century, were in fact electric. In 1900, some 65 years after they were invented, electric cars outsold all other cars in the United States. They had a top speed of 20mph. Electric cars disappeared in the early twentieth century only when cheap oil and better roads meant people could drive longer distances.

Zermatt's electric vehicles were specially made by a Swiss company and can take up to eight people plus luggage. They may look a bit toy-town, but the people of Zermatt accept them because they produce no pollution and they make no noise. The electricity also comes mostly from hydro stations. Year after year Zermatt has voted to keep the ban on private cars. The townspeople apparently understand that the best way to make their village attractive, healthy and safe is to take as much traffic as possible off its roads.

Toy town taxi – there are no private cars in Zermatt, and all delivery vehicles and taxis are electric and the same shape.

batteries that can be swapped out and charged at home. But the problem of the carbon intensity of grid electricity will remain, whatever happens to batteries.

Air quality

People sometimes say diesel is better than petrol in terms of carbon emissions. It is, but only by a tiny amount (5 per cent). It's up to 17 times worse in terms of air quality. According to two of the country's leading experts on air quality, Prof, Frank Kelly of Kings College London and Simon Birkett of the Campaign for Clean Air in London, diesel fumes are basically sending us all to an early grave and very few of us know anything about it.[15]

It's the equivalent of a lifetime of heavy smoking. For our urban children it means their lungs never fully develop. But the latest evidence is that it doesn't just lead to lung diseases like asthma. Diesel soot particles are so minute that they get into our bloodstream and cause heart diseases. And for those with diabetes pollution seems to worsen their ailment.

The message is clear – we need to stop using diesel. Electricity sourced from renewables and biomethane from food waste (see Chapter 6, page 109) have got to be the way forward. We also need to create more greenery, since this soaks up CO_2 and other pollutants.

» *Raise awareness about the diesel problem, lobby the council's Transport Policy Department and ward councillors for monitoring stations in traffic hotspots, and try to get the data published in real time in a way the public can understand*

Electric-car clubs in London, San Francisco and Paris

In 2009 the London Borough of Westminster and the city of San Francisco both launched electric-car-club pilots. Paris is working on a plan to replicate its comprehensive free bike scheme (see page 177) using electric cars. For the foreseeable future I think this is where we will achieve the best fit between electric cars and those who do not have off-street parking in urban areas.

Car-club cars go back to the same spot, so they are ideal for combining with electric charging points. And on average they spend 60 per cent of a 24-hour period in their parking place, so they should always be charged. The car clubs are working on technology that would allow them to see how much charge the car has in it when it gets back to its stand. If the charge is low, then car-club members will be directed to another vehicle.

How can we get more people cycling?

Cycling has increased in Britain in recent years, partly as a response to congestion and crowded public transport, and partly, in London, as a response to terrorism. But we're still not at critical mass such that the hordes of cyclists naturally slow down the traffic. In urban areas cycle lanes are usually part of the main traffic flow and often disappear altogether, or are blocked by parked cars, or made dangerous by speed bumps. In the countryside cyclists often find themselves on large roads or forced to make huge detours on tiny, and often vertiginous, country lanes.

Liability

In the Netherlands motorists are made responsible for any accidents involving cyclists or pedestrians. It seems unlikely that a British government would change the law that radically, so the goal for a local council should be to introduce safety measures to increase cycling among those who do not currently cycle and who are nervous about braving busy roads. As well as the obvious benefit to both the health and fitness of individuals it will also benefit the community by reducing demands on public transport and usage of vehicles for short journeys and cutting carbon emissions.

Free bike schemes

A 'free' bike scheme is being set up in South Camden/ Westminster/the City of London similar to the ones installed in Paris (see box) and several other European cities. My worry is that we will put in the free bikes but we will not make it safer to cycle. One of the keys to the success of the Paris 'free' bike scheme was the massive extension of cycling infrastructure throughout the city. Most UK cycling infrastructure is painted on the roads in the form of bike lanes (that are mainly in the gutter), bicycle icons and advanced stop lines (ASLs), which are also known as bike boxes. There is no enforcement because cycle lanes are advisory and the police say they do not have the time to enforce the £30 fine for motorists who enter the bike boxes.

Battling trucks in the bike box – a typical day in the life of a London cyclist.

If cyclists are to feel secure on busy roads, then someone has to enforce cyclist rights. Only the police can do this. The quid pro quo is that police also act to

Taking a political risk on cycling in Paris

In July 2007 the Mayor of Paris, Bertrand Delanoe, launched a plan to transform Paris into a cycling city. At that point cyclists (and cycle ownership) were virtually unheard of in the French capital. He authorised the installation of a massive free bike scheme called Velib (free bike), something that had already been done successfully in Lyon and La Rochelle. By the end of 2007 there were 20,000 bikes on 1,500 bike stands throughout Paris. Equally important was the creation of an extensive cycling infrastructure: shared pavements, where they were wide enough; extended pavements for raised cycle lanes; and separated cycle lanes in the road, often at the cost of parking bays.

Velib users register a credit card as a deposit guarantee and are then given a code that allows them to remove bikes from stands. They have 30 minutes' free use. After that charges are applied (1 euro for the second half hour, 2 euros for the third half hour, etc.). If a Velib station has no bikes or no space to return a bike, then an interactive map shows the locations of the nearest free bikes or spaces. Problems with checking in or checking out bikes can be resolved by calling the 24-hour Velib switchboard via an intercom at any of the bike stations.

There is roughly one Velib bike station every 300m in the city. The scheme has also had two significant secondary effects – it's encouraged more private bikes on to the streets and, critically in a city of crazy drivers, it's forced motorists to become aware of cyclists.

It just goes to show what politicians can do when they put their minds to it. Turning Paris into a cycling city was a political risk for Delanoe, but he took it anyway. Some 22,000 users on day one of the Velib did not mean the motor car had been dethroned and the battle against climate change had been won, but it was a big step forward.

stop bad cyclist behaviour such as cycling on pavements, cycling without lights, cycling the wrong way up one-way streets and jumping red lights.

Ask your local Police Safer Neighbourhoods Team for help to enforce cyclist rights and to penalise antisocial cycling in your area

London's new Cycle Super Highways

In July 2010 the Mayor of London unveiled his new cycle superhighways along the side of busy commuter routes. I'm sceptical. They basically amount to a strip of blue paint on very busy and polluted roads. And what happens when the cycle lanes hit a bus stop? They simply disappear and cyclists are left to do what they always have to do – fight with traffic.

Separated cycle lanes

Separated cycle lanes are surely the key to coaxing newcomers on to bikes. Experienced cyclists are conflicted about separated cycle lanes, arguing that they coop cyclists up and may be more dangerous, particularly at junctions. The biggest-ever review of separated cycle lanes, published in Denmark in 2007, showed that separated cycle lanes materially increase cyclist volumes, but recommended bringing cyclists back into the stream of traffic before major junctions. Where this was not done accident rates went up because drivers were less aware of cyclists.

In London, where there are separated cycle lanes, they're often over-engineered and, where the cycle lanes are both on the same side of the road, they're frankly dangerous because they confuse motorists who don't expect cyclists to be going in both directions on one side of the road. Two-way separated cycle

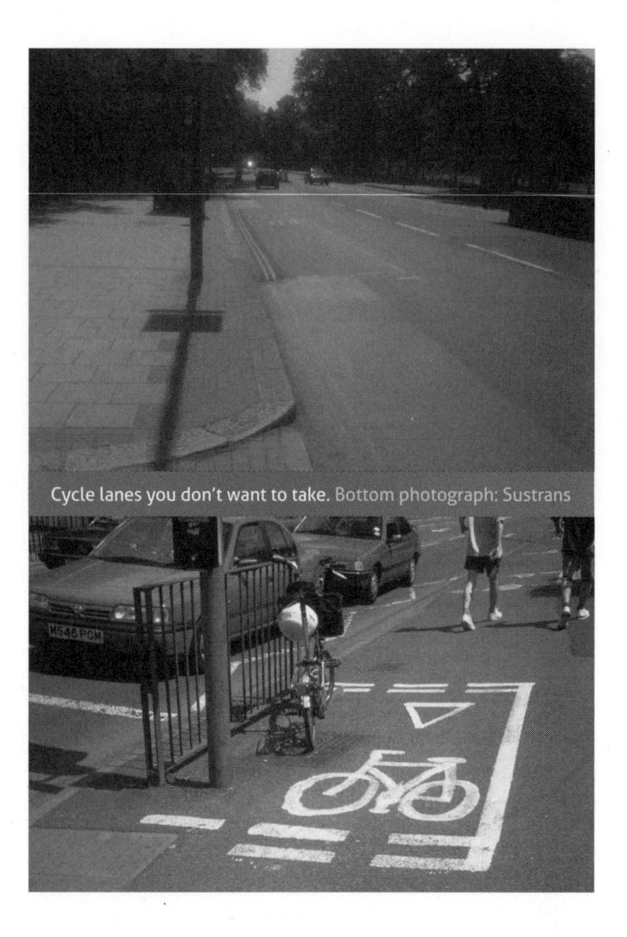

Cycle lanes you don't want to take. Bottom photograph: Sustrans

lanes can also lead to cyclist collisions, as the numbers using the lanes increase.

I would like to see a significant extension of cycling infrastructure throughout towns and cities, with cycle lanes on the same side of the road as the traffic flow. The preference should be to extend the pavement out wherever possible and put a clearly marked cycle lane with minimal separation from the pedestrian space. The second-best option would be to move a line of parking out slightly and put the cycle lane between the pavement and the parked cars.

Women want separated cycle lanes

On 19 January 2010 Sustrans took a 'Motion for Women' petition to Westminster, Edinburgh and Cardiff. It had been signed by 9,000 people from across the UK and was delivered – by bike, naturally – to the House of Commons and formally presented to Sadiq Khan MP, the Minister for Transport.

Meanwhile in Cardiff, Rosemary Butler, the National Assembly for Wales's Deputy Presiding Officer, who learned to cycle in 2009 with help from Sustrans Cymru, received the petition on the steps of The Senedd. And Scotland's Transport Minister, Stewart Stevenson, accepted the petition outside the Scottish Government building in Edinburgh.

The petition – backed by organisations including Mind, the National Federation of Women's Institutes, the Townswomen's Guilds, Women in Rural Enterprise and Zest magazine – was launched in September 2009 in response to research from Sustrans that showed a staggering 79 per cent of women in Britain never cycle at all. The most common reason women gave for not cycling was not feeling safe enough. 67 per cent of women said that cycle lanes entirely separated from other traffic would encourage more women to cycle.[16]

Sustrans, 2010

Women's rights – women cyclists calling for properly separated cycle lanes. Photograph: Sustrans

Schools in the state or private sector that provide comprehensive, verifiable and sustainable travel plans, and who identify black spots, e.g. difficult roads to cross or dangerous roads for cyclists, should be rewarded with engineering assistance towards making key school routes safer. One particular way to reward a school that completes a travel plan would be to install separated cycle lanes along the roads leading to that school.

All of this requires long-term lobbying of ward councillors, schools, parents, Transport Policy Officers and the council's Executive or Cabinet. As with car exclusion zones, if community groups can reach out to schools first, and come up with a vision, then going to the council becomes much easier.

» *Work with local schools and cycling groups on where separated cycle lanes should go so as to best facilitate travel to school by bike, then get parents on board, then go to the council's Transport Policy Department*

More training of drivers

If we continue to allow trucks into our towns and cities we need to find ways to train drivers to be more aware of cyclists. The London Borough of Lambeth has led the way on this, working with its rubbish contractor Veolia to train up drivers to watch out for cyclists.

» *Ask the Transport Policy Department to train the council's drivers to be aware of cyclists*

Health and congestion benefits of cycling

A 2007 report by Cycling England[17] calculated that an adult who opted to undertake a return journey of 2.5 miles (the average cycle trip) by bicycle rather than by car produced savings worth £137.28 through the benefits of reduced traffic congestion, while a regular cyclist saved the NHS £28.30 a year through health benefits. The study concluded that, by making a £70m annual investment in cycling initiatives, the government could reduce car journeys by up to 54 miles a year by 2012 and carbon emissions by 35,000 tonnes. The problem for a local authority is that the financial benefits (reduced congestion, reduced NHS expenditure) do not accrue to the borough.

Communities and councils could open discussions with their Primary Care Trusts to discuss joint funding for cycling infrastructure as a proactive health measure

What can we do to decarbonise freight?

Frankly, trucks should simply be banned from urban areas. They're a source of significant carbon emissions, a menace to pedestrians and cyclists, and they spew out diesel fumes, which are sending us all to an early grave. All the good work Transition Belsize has done with Budgens in Belsize Park feels undone when juggernauts hove into view along my road, shaking up every house as it hits the speed bumps and throwing up clouds of pollution, and then struggles to reverse into the back entrance of the local supermarket, which is in a small residential road.

For towns and cities it should be worth exploring freight consolidation so that freight can be decanted into smaller, more environmentally friendly vans before it enters a town or city centre. This is what happens in the Swiss town of Zermatt (see page 175).

We should also reinstate our canals as freight routes. Walking along Regent's Canal in Camden recently I came across four men pulling a canal boat laden with building supplies. The engine had developed a fault so they were reviving an old skill – except using human-power rather than horsepower!

Suggest a community-run visioning workshop to your council's Transport Policy Department or the Local Strategic Partnership to look at how freight could be managed more sustainably, and invite key players such as British Waterways to participate

Can communities and councils do anything about aviation emissions?

According to the Tyndale Centre for Climate Change, if current trends continue, then aviation will be responsible for more than 100 per cent of the UK government's carbon emissions target by 2050. But CO_2 is only part of the aviation problem. Jet aircraft produce water vapour trails, which act as greenhouse gases. To calculate the overall effect on climate change of an aeroplane you have to multiply the total CO_2 by 2.7. Taking this into account, aviation has the capacity to account for more than 400 per cent of our permissible emissions by 2050![18]

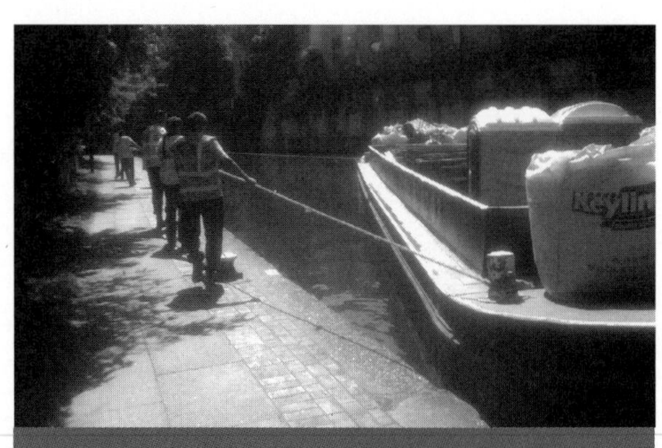

Back to the future – pulling a barge along Regent's Canal in Camden.

Of course the rising price of kerosene may ultimately solve the problem of aviation emissions, but the aviation industry (and Lord Adair Turner, who chairs the UK's Climate Change Committee) talk optimistically about flying planes on 'sustainable biofuels' that don't freeze at altitude. Maybe this will work one day but for the moment it looks like an optimistic long-term techno-fix and in any case it won't deal with the problem of water vapour trails, which create more greenhouse gases than the CO_2 emissions derived from burning kerosene.

Expanding airports, as the Labour government that lost the 2010 general election was trying to do, is a complete nonsense for a country trying to reduce its emissions.

Inconsistent government policies

But perhaps the biggest problem with airport expansion (and road building) is that it sends the wrong messages. The UK was the first country to bring in a Climate Act requiring the government to reduce its emissions year on year. But the proposed expansion of aviation would have destroyed any chance of the UK hitting its target of 80 per cent emissions cuts by 2050. It would be much more sensible to put public money into railways and video-conferencing technology.

» *If airport expansion is proposed for your area, then remember that councils and community groups can work together through the courts to reverse such decisions (see box)*

Aviation policy overturned

In March 2010 the High Court delivered an amazing verdict that left the Labour government's Heathrow policy in tatters. It ruled that the decision by ministers to give a green light to the proposed third runway did not hold any weight. Lord Justice Carnwath dismissed the government's claims to the contrary as "untenable in law and common sense", and said that if the government wanted to pursue its plans for Heathrow expansion, it would have to go back to square one and reconsider the entire case for the runway.

The implications of the ruling are profound, not just for Heathrow but for airport expansion plans across the UK. Lord Justice Carnwath ruled that the 2003 Air Transport White Paper – the foundation of expansion plans across the country – was obsolete because it was inconsistent with the Climate Change Act 2008. The judge ruled that the government's entire aviation policy must now be reviewed to take into account the implications of the 2008 Climate Change Act. He found that "the claimants' submissions add up, in my view, to a powerful demonstration of the potential significance of developments in climate change policy since the 2003 Air Transport White Paper. They are clearly matters which will need to be taken into account under the new Airports National Policy Statement."

The challenge was brought by Hammersmith and Fulham, Hounslow, Hillingdon, Richmond upon Thames, Wandsworth and Windsor and Maidenhead councils, with support from Kensington and Chelsea, Transport for London and the Mayor of London. They were backed by the local residents group (Notrag), aircraft noise campaigners HACAN, the World Wildlife Fund UK, the Campaign to Protect Rural England and Greenpeace. The Royal Society for the Protection of Birds was an expert witness.

This was a famous victory, and, assuming the decision isn't overturned on appeal, it seems to me to illustrate two things: 1) how much the situation has changed in recent years that the courts can ride to the defence of the planet and eco-activists; and 2) that councils and communities can have tremendous influence whenever they work together, even on aviation policy.

The coalition government that came to power in the UK in May 2010 immediately threw out plans to expand Heathrow, Gatwick and Stansted. However, it didn't rule out expansion elsewhere, so campaigners and councils need to be on their guard and be prepared to use the Carnwath Ruling if need be.

Chapter 12

WATER

Questions answered in this chapter:

- Why is water such a difficult issue?
- What are the factors in the coming water crisis and what can we do about it?
- What can you do to protect your community from water risk?
- How can communities collect as much of their own water as possible?
- What rules need to change to create incentives for action on water?
- How can we celebrate water more often?

Although the main focus of this book is cheap fossil fuels, it makes little sense to consider carbon in isolation as all too many people do – it's simply the most worrying of the natural resources we humans are using up because its looming scarcity could provoke the breakdown of society in the industrialised West, and it has the capacity to provoke runaway climate change.

Yet the situation with water is in many ways just as critical. The difference between carbon and water is that we have come to depend on the former but we absolutely have to have the latter. However, fresh water is a natural resource under immense and growing pressure from human society because of the traditional pressures of development and the new pressures of climate change and peak oil. Very few people have yet started to comprehend the scale of the water problems that are coming. It's been hard enough to get to grips with carbon, but water is a lot more complicated, primarily because of the difficulty of who owns it and how it should be priced.

Why is water such a difficult issue?

Most fresh water is free (rainfall) or virtually free (abstracted water) because it's either subsidised by governments (e.g. for agriculture) or because its price doesn't account for externalities such as scarcity, pollution, salinity, soil erosion and flood damage.[1] As a result, water has historically been consumed as if it were a permanently abundant resource.

It would be much easier to price scarcity and environmental damage if the issue of ownership were settled. In the UK the Environment Agency is responsible for licensing the extraction of groundwater and the return of waste water to the environment, but not rainwater.

In Germany, to prevent flooding, some municipalities effectively enforce ownership of rainwater by imposing water run-off taxes to households and businesses if they allow rainwater to flow off their property (see page 191). In California the system is 'first in last out', so the earliest settlers have a prior claim *whatever* they are doing with the water! In Australia's Murray River basin, which is in the grip of a multi-year drought, the government allocates water.[2] There is no single or clear way to decide water rights.

The main problem I identified at Camden Council was very simple – nobody was responsible for water. Indeed when I first mooted the idea of writing a Water Report the officer advice internally was: "this one won't be a problem – we don't use much water"! That conclusion was reflected back to me elsewhere. Alex Inman, a consultant for the Westcountry Rivers Trust, a charity seeking to improve water quality in its area, said: "Local councils don't seem to have anyone who has the interest or the remit to want to work with community groups on water."

However, I think it's also fair to say that very few community groups seem to be working from the bottom up on water. Almost everything about water management is a top-down process – metering, flood prevention, planning rules, water run-off targets, etc. As yet I haven't found one Transition Initiative in the UK that's created a water group, although there are several in Australia, perhaps unsurprisingly given the problems out there.

Maybe the lack of action on water in the UK stems from a sense that so many other things seem more urgent. Or perhaps it's because most people think of the UK as a wet country, which of course some of it (the north-west) is. But the south-east is a designated water-stressed area, with less water available per person in terms of rainwater, river water and groundwater than Sudan or Syria. London has less available water per person than Istanbul or Jerusalem.[3] This will only get worse as climate change kicks in. By the 2040s the exceptionally hot summer of 2003, which turned our parks and gardens into deserts, is expected to be the norm.[4]

The 2008 Intergovernmental Panel on Climate Change technical report on climate change and water says: "Many semi-arid and arid areas (e.g. the Mediterranean basin, western USA, southern Africa and north-eastern Brazil) are particularly exposed to the impacts of climate change and are projected to suffer a decrease of water resources due to climate change [high confidence]".[5] These are all areas that export food to the UK.

In other parts of the world rainfall is set to increase, with hurricanes and flooding making farming less predictable. "The frequency of heavy precipitation events (or proportion of total rainfall from heavy falls) will be very likely to increase over most areas during the 21st century, with consequences for the risk of rain-generated floods".[6]

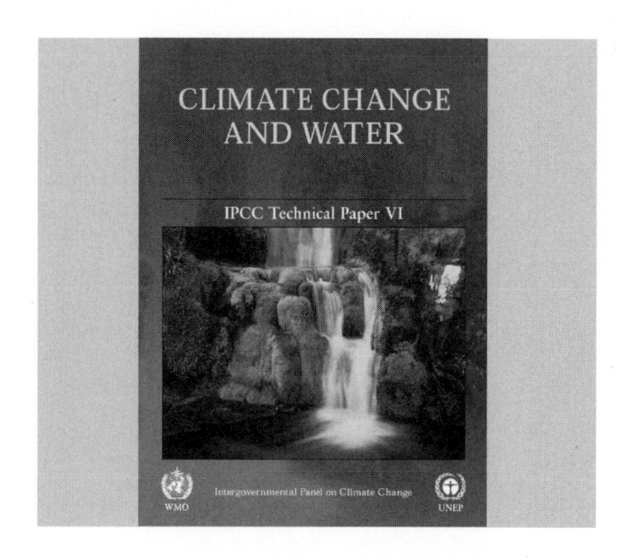

Rising sea levels are causing salination of coastal farms, especially where aquifers are exhausted or rivers are running dry. Over-abstraction further inland is causing dryland salinity.

Farming is set to become more precarious everywhere. There may well come a point when food supply lines break down because of a catastrophic reduction in available water (drought, over-abstraction, pollution, increased demand) or because of excessive water (flooding, hurricanes) or the wrong kind of water (salinated, polluted). The result is a greater threat to fresh water supplies and food supply chains than the human race has known outside wartime.

We need to start thinking much harder about water. And perhaps more than on any other issue communities and councils need to stop seeing it as someone else's problem and take more responsibility for it.

What are the factors in the coming water crisis and what can we do about it?

Demographics, development and pollution have traditionally been the concerns in relation to stresses on fresh water supplies. The world population is growing rapidly. From six billion today it is predicted to reach nine billion by 2050. According to the UN, 29 per cent of the global population lived in urban regions in 1950 and an estimated 60 per cent – 5 billion people – will be in cities by 2030.[7] The world is on the move and is heading mainly for water-stressed cities in coastal regions.

As populations grow richer they use more water. US water demand has tripled in 30 years whereas the population has increased by only 50 per cent.[8] Richer populations demand more meat and dairy. which require huge amounts of water to produce. To provide 500 calories in the form of corn requires 130 litres of water, while the same number of calories produced as beef requires 4,900 litres because the grain to feed the cows has to be grown, then the cows have to be given water to drink throughout their lifetimes.[9]

The water used to create a product is called embedded or virtual water. The daily food intake of the average UK citizen requires 3,400 litres of embedded water as it goes from farm to plate.[10] Americans consume twice as much – 6,800 litres of virtual water per person every day.[11] The average hamburger contains 1,400 litres of embedded water. A cup of coffee requires 140 litres of water to grow, produce, package and ship the beans – roughly the same amount of water used by an average person daily in England for drinking and household needs.[12]

» *Raise awareness about the problem of embedded water in food (check out the Footprint Trust and Waterwise for info)*

Another aspect of development and population growth has been the industrialisation of farming and the use of chemicals to increase productivity – the so-called 'green revolution', which turned out to be anything but green. Here I've found pockets of hope. For example, the Westcountry Rivers Trust is a community group that's been working hard with its local farmers to reduce the toxicity of water running off farming land (see box, opposite, top).

» *Work with your local farmers on improving the quality of water running off their land*

The increasing carbon cost of water

Water that's stored, pumped or cleaned has both a financial cost and increasingly a carbon cost as well. Building reservoirs, pumping water around countries, cleaning water, treating water, irrigating fields,

WESTCOUNTRY RIVERS TRUST – WORKING WITH FARMERS ON WATER RUN-OFF QUALITY

Alex Inman, Consultant, Westcountry Rivers Trust

This is a rural area, full of farming, and there have historically been big problems with water pollution from nitrates and phosphates getting into the soil. The Trust has been working with farmers on a very confidential basis. We've had money from the EU. DEFRA's helped a bit. It's about helping farmers to understand how you use pesticides and when. It's non-governmental brokering and encouraging and empowering, not using legislation.

Carrots are win-win solutions. We work with the farmers, we work out with them ways they can save money and improve the environment. So, for example, quite a lot of farmers use too much fertiliser. Grass just doesn't need 350kg of nitrogen per hectare, it only needs about 220kg for it to grow optimally. We provide the technical expertise, working with about 3,500 farmers. Most of them call the Trust on a regular basis. We promoted what we call 'extensive agriculture', which is using as few artificial inputs as possible.

It's a question of the stage the farmer is at, it's a continuum, if you've got people who behave in one way, and there's an awful long way for them to go to start behaving in another way, you can't expect them to jump straight to A B C and D straight away. It's moving in the right direction, so the argument would be, if you've got lots of people in society moving in the right direction, moving a little bit, then that's much better than having a few people doing a lot.

See www.wrt.org.uk

washing agricultural products – these require fossil fuels or electricity derived from fossil fuels. As the price of oil rises because of peak oil, the underlying cost of supplying water will increase as well.

The largest-ever spike in demand for electricity was at the end of the England–Germany penalty shoot-out in the 1990 World Cup semi-finals when a million or so kettles were switched on.[13] But the water companies also had a huge spike in demand, partly because those kettles had to be filled, but also because a million toilets were flushed at the same time. That meant a lot of water had to be pumped.

Raise awareness about the carbon cost of water

Water efficiency and grey-water recycling

We do a crazy thing in the UK – we clean water to drinking-quality standard and then use it to wash our windows, flush our loos, clean our cars and water our gardens. In Germany many homes and businesses have a dual pipe system – one for precious clean water, which is used only for drinking, cooking and

Water Works on the Isle of Wight

The Isle of Wight imports some 30 per cent of its water from mainland Britain, which puts a particular strain on natural resources. In response to this the Footprint Trust, an educational charity, launched an initiative in 2006 called Water Works, which raises awareness about the need to use water resources more efficiently.

The Trust works with the community, the council, businesses and other organisations to reduce waste by promoting water efficiency in gardens and in the home. The charity attends public events on the Isle of Wight to promote water efficiency and sustainable living. It has also created demonstration drought-tolerant gardens in public places.

See www.footprint-trust.co.uk/water_works.htm

Cleaning grey water with small and large reed-bed systems

Reed beds are a natural habitat found in floodplains, waterlogged depressions and estuaries. Water trickling through the reed bed is cleaned by micro-organisms living on the root system.

Chris Shirley is a London-based inventor who's patented a small-scale grey-water cleaning system that can be installed on a roof or in a back garden. His Green Roof Water Recycling System, GROW, is essentially a tiered garden of low-growing, flowering, native plants, whose roots are supposed to perform the same cleansing function as a reed bed.

The Centre for Alternative Technology (CAT) in Machynlleth in Wales has a centrifugal-force compost toilet with reed-bed cleaning. After leaving the toilet block the solids and liquids are separated by circling round something called an Aquatron. This device, shaped like an hourglass, and visible to CAT visitors, uses simple principles of surface tension and centrifugal force to separate toilet solids and liquids. The solids are turned into compost. The liquids are cleaned through a reed-bed system and returned to nature.

Reed-bed systems can be large as well as small. The Welcome Break Motorway Services Area on the M40 near Oxford, which opened in July 1998, was built with a large waste stabilisation pond and reed-bed system designed to treat all waste water on site.

Dealing with human waste – creating reed beds at the Centre for Alternative Technology in Wales to treat the liquid outflow of the centre's toilets. Photograph: CAT

washing, and the other – grey water – which is used for everything else.

In a grey-water recycling system rainwater and waste water from bathrooms and kitchens are collected and filtered in an underground tank, then recycled through toilets and on to gardens. It will rarely make sense to retrofit grey-water recycling systems on existing houses and flats – in 2009 it cost approximately £8,000 to retrofit a Victorian house.[14] Unless and until the price of water goes up, grey-water recycling systems will make real sense only for large organisations and for new build.

» *If there's a site in your area that you think might benefit from a reed-bed system, then contact the council's Sustainability Team and offer to set up a site visit with a reed-bed specialist*

I've had numerous plumbers round over the years and explained to them that I wanted to divert my bathroom waste water on to the garden because of the hosepipe ban. They all scratch their heads and eventually say it can't be done or that they don't know how to do it. And yet it's so simple.

Grey-water diverters to funnel bath or shower waste water on to the garden are relatively easy to install, cost very little (£30) and work a treat so long as you have a little gravity on your side, i.e. your bathroom is on at least the first floor and not the ground floor.

Councils could coordinate the establishment of training seminars for plumbers/installers at local colleges or other educational facilities for rainwater harvesting or other environmental technologies. This is partly about joined-up thinking – if planning committees are requiring rainwater harvesting and grey-water recycling systems in new developments, they should also be looking at improving the skills of those who end up having to do the work. This goes back to a constant refrain – that councils should act as enablers wherever possible. I'm sure residents would do more to save water and energy if they knew what to do and whom to contact.

» *Lobby your council to work with local technical training establishments to teach plumbers about water-saving devices*

The Bathroom Manufacturers Association can now supply a long list of devices that save water, from taps that make an audible click sound as you turn up the water flow rate to bath tubs that mould to your body! (See Resources section.) Transition Town Totnes has also put together an extremely useful workbook, *Transition Together*, which includes advice on cutting water use – contact the Transition Network for details.

Councils can set tough standards for water fixtures and fittings for their buildings, and really must do so in water-stressed areas such as the south-east of England. Social housing can be upgraded opportunistically on change of occupant or when an estate is refurbished. Councils can also introduce tough standards for water fixtures and fittings in Local Development Frameworks.

» *Lobby the council's Housing and Property Services Departments to fit water-efficient devices in their buildings*

My favourite water-saving device is the waterless urinal, which I first saw at Nottinghamshire County Council. They basically do away with the trap under the urinal and pipe the urine straight out of the building instead. That means the urinal simply needs a wipe down now and again. In traditional gents' loos

No need to flush - a waterless urinal saves thousands of litres of water. Photograph: Centre for Alternative Technology

urine collects in the trap and smells, which is the reason why they need constant flushing. Apparently waterless urinals save huge amounts of water and money.

Revolving Energy (or Invest to Save) Funds are well established in local government as a way of identifying energy efficiency measures that could save money. Why not do the same for water? The evidence from those who have installed water-efficiency measures like, say, waterless urinals, shows that they pay for themselves. In other words councils could put up the money, safe in the knowledge that it will come back in time.

» *Suggest to your local pub that it installs waterless urinals as a way of saving money and to your council's Property Services Department that it create a Revolving Water Fund*

What can you do to protect your community from water risk?

Most people now know that global warming doesn't simply mean the planet will heat up; it means the

Flooding farmers' fields

The West Country Rivers Trust has been working on a project where large businesses would pay farmers to effectively flood their land in times of heavy rains and provide part of a flood-mitigation system. Alex Inman, formerly of the Trust, told me: "We're looking at working with insurance companies to help finance these things. So the businesses pay their insurance, then insurance companies pay the farmers. The figures are very convincing. We've had lots of positive feedback, but it's early days."

Deliberate flooding of fields could help in mitigating flood damage. Photograph: iStock

amount of energy in the climate system will rise, leading to increasingly violent weather conditions. So, for example, although overall UK temperatures are predicted to increase, in winter we will be subjected to more violent storms and more rainfall over shorter timescales.

Flooding in the UK in the summer of 2007 cost in the region of £4 billion. Ironically, whilst more than 250,000 people across central and southern England were left without any clean water because of flooding, elsewhere in south-east millions of people were subject to a hosepipe ban because of drought.[15] Flood damage is bound to get worse as climate change brings more powerful storms and because we've built too many settlements on or near floodplains.

Low Carbon West Oxford was set up by residents in November 2007 after flooding hit their area badly and drove home to them the potential impact of climate change around the world. They describe themselves as "a community working together to combat climate change by cutting our individual and community carbon dioxide emissions, and promoting more sustainable lifestyles."

One thing that most definitely could be done is to rebuild flood-damaged buildings not as they were but as waterproof as possible. That means councils and community groups working with local people to explain why it makes sense to prepare for the next flooding event, and working with insurance companies to make water protection measures the norm in any refurbishment of water-damaged properties.

See if there's a local group working on flooding issues in your area and if there isn't, set one up

Sustainable Urban Drainage Systems (SUDS)

Urban sewers already cannot cope with the amount of storm water that falls. That's largely because we've spent the last 50 years concreting over everything in sight. As a result rainwater is pushed as fast as possible to the sewers, which are overwhelmed and so flooding occurs.

We need to replace as many hard surfaces as possible with soft (permeable) surfaces. Highways Departments need to use permeable paving stones for their repair work from now on. Ideally, we need to green as many hard surfaces as possible.

For example, front gardens and driveways do not need to be paved – they can be permeable and still provide parking facilities. Green roofs soak up storm water, which is another way of preventing flooding. But we also need to install waterscapes wherever possible with swales that attenuate storm water. This is common place in Germany, where flooding is, as in the UK, a growing problem. A swale could be something like a football pitch that is built in a slight dip. When the storm hits the pitch it becomes a swimming pool. When the storm is over, the water is then released slowly into the water table or the sewers.

 *Ask your council if it has a plan for flooding and whether it has considered subsidising green roofs or creating swales, and whether SUDS are required in planning applications*

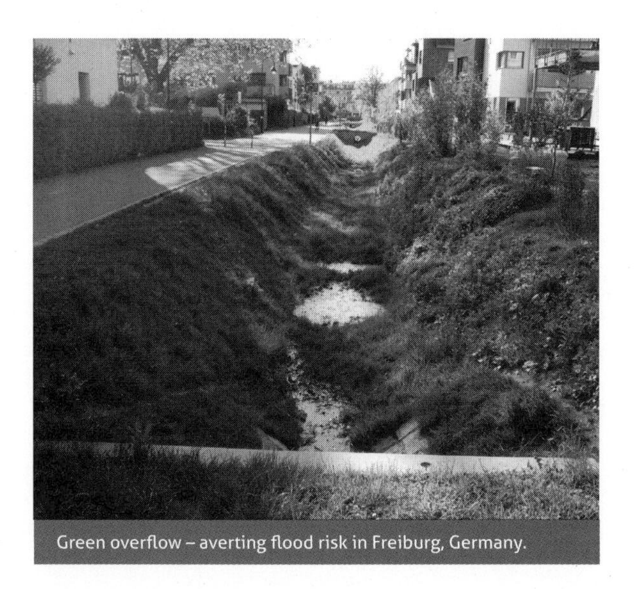

Green overflow – averting flood risk in Freiburg, Germany.

Flood defences and civil contingencies planning

London's main defence is the Thames Barrier. Up to February 2009 the Thames Barrier had had 114 flood-defence closures since its completion in October 1982, most of them since 2000. Just across the North Sea Rotterdam lies behind several hundred kilometres of dykes. A network of sluices, locks and barriers give it some of the world's best sea defences, which would be breached only by a one-in-10,000-year storm. But the risk remains, and will grow over time, that a major flood could breach the defence system, affecting around half the Dutch population.

That's why Rotterdam is busy installing underground water-storage facilities. A multi-storey car park being built near the city's research hospital will cater for 1,200 cars and 10,000 cubic metres of water, which will be stored in a reservoir at the base of the structure. The city council is also subsidising the installation of green roofs to hold up storm water and exploring the concept of 'water plazas' that can be used as playgrounds during normal weather conditions and as swales to collect water during torrential rain.

Flooding is a civil contingencies issue. Councils are already represented on regional civil contingencies planning groups. Community groups such as Transition Initiatives should be too. (See Chapter 2, page 45 for more on civil contingency planning.)

 *Ask your council's Emergency Planning Officer to help to get representation for your community group on the regional civil contingencies planning body*

How can communities collect as much of their own water as possible?

The total amount of water on the planet never changes – it's endlessly recycled (rain, land, river, sea, evaporation, clouds, rain . . .). This leads some to conclude that water is a renewable source, but fresh water is most definitely not a renewable resource. We waste it, we pollute it and we flush it out to sea. It takes resources to renew it, usually carbon resources, or enough natural capital that the biosphere can clean it. We're running out of cheap energy and the biosphere has been degraded so much that it is losing its capacity to create clean water, so the best thing you can do, after using less of it, is to find your own local supply.

Groundwater supply

Underneath Camden there's a water supply (the water table or groundwater) that's accessible via boreholes. Two of Camden's swimming pools are filled with water in this way, but in urban areas there's generally little scope for individuals or community groups to use boreholes. You can apply to the Environment Agency for a licence to make a borehole for your property or you can do it through a planning application to your local council, but this is likely to become increasingly difficult in water-stressed areas.

Rainwater harvesting

We need to collect and store a lot more water from the skies. That means creating tanks for water storage and it means finding ways to keep the water clean. The easiest thing to do if you have access to a drainpipe is to install a water butt. Something like 85,000 litres of rainfall falls on the average UK roof each year, so there's plenty to collect. I have a water butt perched on my back balcony at the first-floor level, which means I almost never have to use tap water on my strawberries, fennel and tomatoes!

Councils use water outdoors in several ways. They water plants and green spaces, and wash buildings

Bottled water

A few words on the utter craziness of bottled water – the very symbol of a society gone materialist and mad. We transport clean water over great distances in glass and plastic bottles in the name of health or taste. But the water from our taps is, in most cases, just as good. Coca Cola famously proved this point inadvertently when it was caught bottling tap water in Croydon and selling it at a huge margin as Dasani mineral water. It's also no better for us. The Consumer Council for Water has a good downloadable document, 'Why tap water is a winner', which explains the facts (www.ccwater.org.uk/upload/pdf/why_tap_water_winner.pdf).

So let's stop selling or providing bottled water and install or re-plumb water fountains in buildings and public places. And buy some jugs! Some restaurants now provide filtered tap water instead of bottled water. That sounds like good practice to be spread among catering establishments.

It's perhaps worth pointing out that there's no legal requirement for restaurants to serve tap water or to force you to buy bottled water. Also, if they do serve tap water, then they can legitimately charge for it because the provision of tap water includes an element of service, such as pouring the water into a jug and/or glass and cleaning the jug and/or glass, so the Supply of Goods & Services Act applies.

In 2007 there was a variety of grassroots and media campaigns trying to persuade people to drink tap water rather than bottled water, and Islington became one of the first councils to ban bottled water on its premises. This seems to me to be fertile territory for cooperation between community groups and councils.

and vehicles. At the moment, ludicrously, most of this is done with drinking water. All councils should be collecting rainwater wherever they can for use on outdoor spaces.

» *Most councils don't have Water Officers, so lobby the council's Sustainability Team, Housing or Parks Departments about collecting rainwater for watering outdoor spaces*

» *Start a campaign in your area to phase out bottled water in restaurants and the council, and to bring back public water fountains*

What rules need to change to create incentives for action on water?

Camden's all-party Sustainability Task Force recommended that a water specialist be appointed to be responsible for water and that that person work in cooperation with the relevant statutory organisations, such as the Environment Agency and Thames Water, to prepare a comprehensive water strategy. The London Water Strategy was published in draft form in September 2009.[16]

Bullet points from the draft London Water Strategy:

- Reduce leakage.
- Install water meters in homes.
- Attain Level 6 of the Code for Sustainable Homes – the highest in new buildings.
- Reduce use of bottled water in favour of tap water.
- Encourage green roofs, rainwater harvesting, grey-water recycling and SUDS.
- Joint working with strategic partners on flood risk.
- Sewage to be used to create biogas and run turbines to create electricity.

There's a lot of scope for putting together a coherent water policy for a local community in conjunction with the Environment Agency, the local water company, the council, major consumers of water and citizens, but I think the council needs a water specialist if it's to be able to participate meaningfully.

» *Lobby your council's Sustainability Team to employ a water specialist, which might come anyway if you can persuade the council to start a Revolving Water Fund (see page 187)*

Metering

DEFRA's 'Future Water' strategy, which was published in February 2008,[17] says: "Our current system of charging based largely on the value of people's homes 35 years ago is archaic and rife with anomalies. We need a fairer system that offers incentives to conserve water. In areas of serious water stress it's pretty clear that this will mean near-universal metering before 2030."

Water meters should be installed as a matter of priority wherever possible and well before 2030. Water companies would rather install meters en masse than piecemeal because they see it as more efficient and cheaper. However, they can require the fitting of meters on change of occupancy, both private and rental. In an area of high turnover such as Camden this implies that we could install meters in up to 20 per cent of homes in any one year.

Meters are usually estimated to make households about 10 per cent more efficient in their use of water.

But they also introduce a problem of regressivity – with meters, rich households pay the same per litre as poor households, whereas without meters rich households pay more because bills are based on property size. Charging everyone the same per litre may be acceptable in some societies, or it may be possible to address through means-tested benefits. But I would like to see a progressive and environmentally friendly billing system, where each person receives a free water allocation – let us call it a human right to water – then additional consumption becomes increasingly expensive. I think that would be the fairest way to handle the equity issue and the best way to deal with the environmental problem.

 *Lobby your MP for government action to speed up the introduction of metering, ideally with a progressive billing system*

Porous pavements – soaking up water between the paving stones in Frickengen, Germany. Photograph: Jessica Read

Run-off taxes

In Germany municipalities have the right to introduce water run-off taxes. This is because of the problem of flooding in many areas. Water run-off taxes are imposed on hard surfaces under a householder's control. A non-porous driveway or a tiled roof would therefore attract a charge whereas a green roof or an unpaved garden would not. Councils in areas prone to flash flooding could try using the Sustainable Communities Act (if it survives – see page 47) to lobby the government for the right to introduce water run-off taxes. The money raised could be spent on installing green surfaces and water-efficiency devices in your area.

If it proves impossible to introduce water run-off taxes, councils should introduce water run-off targets for new or refurbished buildings and include water run-off targets as a general principle in their Local Development Frameworks. They could also consider financial incentives for householders to green hard surfaces under their control. In 2008 the government changed the definition of permitted development rights to prevent the paving over of gardens that was happening in urban areas as a reaction to parking pressures. Now we need to reverse the process and start re-greening hard surfaces wherever possible.

 *Lobby your council's Planning Department to introduce water run-off targets and your council's Transport Department to offer an eco-grant and free membership of a car club to anyone willing to un-pave their front garden, install a porous driveway or fit a green roof*

How can we celebrate water more often?

Water is essential to human life, but it can also be a significant source of human happiness. We should do more to celebrate it. The opening of the South Bank walkway along the Thames from Tower Bridge to the London Eye has been a huge success. Apparently it's the most visited part of London by Londoners (as opposed to the most visited part of London, which is Piccadilly Circus). That doesn't surprise me – people like waterscapes.

So popular have natural water features been in places like Germany, and so useful in terms of mitigating flooding caused by climate change and enhancing quality of life, water efficiency, storm-water attenuation and fire-fighting capacity that it seems to me self-evident that we should do the same wherever possible. For example, every school should be designed so that visible water is built into its conception, from internal pipes to external waterscapes.

» *Lobby your council's Planning Department to install, or require developers to install, visible water features, based on natural landscaping rather than traditional fountains*

» *Set up a water festival in your area and create an annual award for water efficiency for businesses and non-governmental organisations*

HAMPSHIRE LEADS THE WAY ON CELEBRATING WATER

Martin Burton, Hampshire Water Partnership

The Hampshire Water Partnership (HWP) was formed in 2001 to implement some of the actions identified in the 'Water in Hampshire – A comprehensive review' report[18] published in March 2000. The HWP is composed of a wide range of organisations who are stakeholders in water in Hampshire. Its aim is to ensure the long-term future of Hampshire's rivers, aquifers and wetlands for people and wildlife, and to foster increased understanding and knowledge-sharing about Hampshire's water environments and the issues facing rivers, wetlands and aquifers. It comprises private, public and voluntary bodies who have a role or interest in the sustainable management of this unique water environment. The Partnership has no executive powers, but provides a forum for closer working relationships, information dissemination, and discussion on topical issues.

The specific aims of the HWP are as follows:

- To work in partnership to develop creative solutions to the range of pressures faced by Hampshire's water environment.
- To represent and promote the interests of partners both to the public and to decision-makers.
- To effect cultural change in how organisations and individuals in Hampshire view the supply and disposal of water.
- To provide a forum to promote the distinctiveness of Hampshire's water environment.
- To disseminate information and advice on water issues to a diverse set of stakeholders across Hampshire.
- To complement organisational statutory powers and best practice.
- To set and monitor measurable targets to implement the Partnership Action Plan.
- To run the annual Hampshire Water Festival, which first took place n August 2003 in Winchester and attracted 10,000 visitors.

See www3.hants.gov.uk/hampshireswater.htm

Chapter 13

WELL-BEING

Questions answered in this chapter:

- Does carbon-fuelled economic growth make us happy?
- Can happiness or well-being be measured?
- Who's doing work on well-being in local government?

Does carbon-fuelled economic growth make us happy?

Economic growth is underpinned by cheap fossil fuels. Without the latter we would not have the former. But burning fossil fuels to create growth is an illusion. Growth on a finite planet cannot be limitless; it is the reason why we stand on the brink of runaway climate change and, perhaps most importantly of all, it doesn't seem to make us happy, at least not in the industrialised world.

Somewhere in the nineteenth century we took a wrong turn. Society forgot the teachings of the liberal philosopher Jeremy Bentham and his acolytes – that the primary aim of governments was to make people happy. Bentham argued that "Nature has placed mankind under the governance of two sovereign masters, pain and pleasure." From this, he derived the founding principle of utilitarianism: good is whatever brings the greatest happiness to the greatest number of people.

Governments today talk about economic growth as if it brings the greatest happiness to the greatest number of people. The Queen's speech at the opening of parliament in November 2009 began with the words: "My Lords and members of the House of Commons. My Government's overriding priority is to ensure sustained growth to deliver a fair and prosperous economy for families and businesses, as the British economy recovers from the global economic downturn."

But, as a society, we're no happier than we were at the end of the 1960s, despite a flood of 'useful' things being produced by our factories and a plethora of 'exciting' financial products being conjured out of thin air by our bankers. In fact, many of us are unhappier. A quarter of the UK population has suffered some sort of mental illness. The accumulation of stuff, the increased travel, the speeding up of life, the endless gadgets, the rupturing of family bonds, the isolation of cities, the endless hours of work to earn money, growth – these have not made people happier.[1]

Walking the dog, growing your own carrots, having friendly neighbours, reading a good book, playing an instrument, feeling loved, not suffering from mental illness – such things as these make us real and happy, not GDP, manufacturing output or inflation targets. Equally, none of those conventional measures of economic progress highlights the damage we are doing to the planet and the risk that we are creating for future generations.

If ever there was a time to stop pretending that economic growth theory is a measure of human progress, it is now. We need to define human progress not in terms of 'stuff' but in terms of what makes us happy.

> "At the end of the day prosperity goes beyond material pleasures. It transcends material concerns. It resides in the quality of our lives and in the health and happiness of our families. It is present in the strength of our relationships and our trust in the community. It is evidenced by our satisfaction at work and our sense of shared meaning and purpose. It hangs on our potential to participate fully in the life of society. Prosperity consists in our ability to flourish as human beings – within the ecological limits of a finite planet. The challenge for our society is to create the conditions under which this is possible. It is the most urgent task of our times."
>
> Tim Jackson, *Prosperity without Growth*[2]

Can happiness or well-being be measured?

Professor Richard Layard wrote the seminal text on happiness in 2005.[3] He showed that you can measure happiness the easy way or the hard way. The easy way is to ask people how happy they are with their lives. The hard way is to also ask someone's family, friends and work colleagues how happy the person is, and then to cross-reference that information with data on bouts of mental illness or unemployment as well as the number of hobbies someone has, and his or her participation in community activities. The really interesting thing is that both methods produce roughly the same result. Either way it's perfectly possible to measure happiness.

How do you make the concept of happiness relevant for the average voter? Well, how about putting it another way – how do you make GDP growth or money-supply figures or rates of industrial production relevant for the average resident? You can't – and anyway, in my opinion, they're not relevant. What is relevant is whether people have a job that makes them happy and earns them enough to support their household; whether they have enough friends and family to feel loved; whether they have hobbies and community interests; whether they suffer from mental or other types of illness. If politicians are to reconnect with society, then they need to re-establish trust, but they also need to understand that making people happy means putting the happiness of people centre stage.

Professor Layard's recommendations for a better world were as follows:

- Monitor the development of happiness in our countries as closely as we monitor the development of income.

- Rethink attitudes to standard issues such as taxes (recognise the role they play in preserving the work-life balance) or employment mobility (reconsider its tendency to increase crime and weaken families and communities).

- Spend more on helping the poor.
- Spend more on tackling the problem of mental illness.
- Introduce more family-friendly practices at work – more flexible hours, more parental leave and easier access to childcare.
- Subsidise activities that promote community life.
- Eliminate high unemployment.
- Fight the constant escalation of wants by prohibiting commercial advertising to children, as in Sweden.
- Improve education, including moral education.

Something positive did come out of Professor Layard's book – the Labour government started employing

Measuring well-being or flourishing

The new economics foundation, nef, has been measuring well-being – or 'flourishing', as it also likes to call it – for years in its National Accounts of Well-being. It calls for governments to "directly and regularly measure people's subjective well-being: their experiences, feelings and perceptions of how their lives are going, as a new way of assessing societal progress."[4]

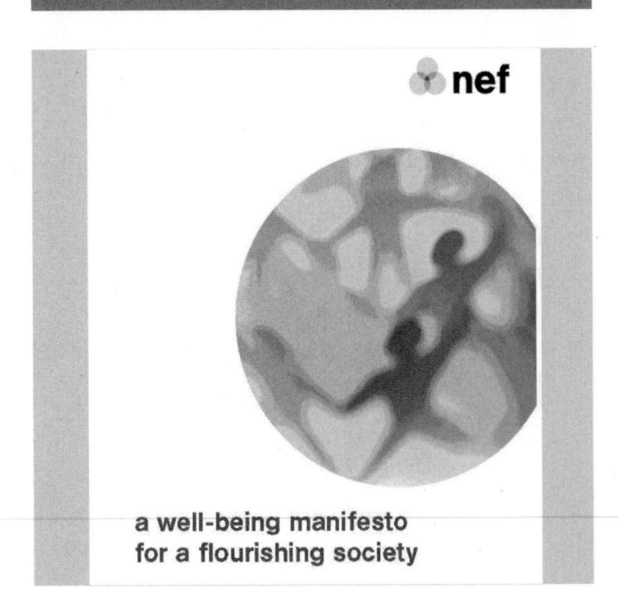

nef's five 'ways to well-being'

1. **Connect** with people around you. With family, friends, colleagues and neighbours. At home, work, school or in your local community. Think of these as the cornerstones of your life and invest time in developing them. Building these connections will support you and enrich you every day.
2. **Be physically active.** Go for a walk or run. Step outside. Cycle. Play a game. Garden. Dance. Exercising makes you feel good. Most importantly, discover a physical activity you enjoy and one that suits your level of mobility and fitness
3. **Take notice.** Be curious. Catch sight of the beautiful. Remark on the unusual. Notice the changing seasons. Savour the moment whether you are walking to work, eating lunch or talking to friends. Be aware of the world around you and what you are feeling.
4. **Keep learning.** Try something new. Rediscover an old interest. Sign up for that course. Take on a different responsibility at work. Set a challenge you will enjoy achieving. Learning new things will make you more confident as well as being fun.
5. **Give.** Do something nice for a friend, or a stranger. Thank someone. Smile. Volunteer your time. Look out, as well as in. Seeing yourself, and your happiness, linked to the wider community can be incredibly rewarding and creates connections with the people around you.

Go foraging not shopping – a happy Saturday morning spent with the Transition Belsize Foraging Group looking for edible plants (and lunch!) on Hampstead Heath.

cognitive therapists to work on making people happy in the here and now, to give them a positive outlook going forward. I'm not enough of an expert to judge the success of this policy, but what I do know is that it didn't stop politicians talking about growth.

Who's doing work on well-being in local government?

Over the last few years a number of UK councils have introduced projects with an element of well-being in them. These include Bristol's Quality of Life survey, Manchester's community guardians, Nottingham's work with young people and Thanet's well-being audit of two wards, which led to the creation of their Safer Stronger Community Fund.[5]

Caerphilly in Wales was the first local authority to introduce a truly comprehensive well-being indicator with a long-term aim of guiding all policy. In 2008/9 it worked with nef on what it calls its 'Living Better, Using Less' indicator, which is made up of life expectancy, life satisfaction and ecological footprint. This tries to work out how happy and healthy residents are – their well-being – and divides this by the amount of resources they use – their ecological footprint. The index is based on the equation:

Living Better, Using Less = (Life Expectancy x Life Satisfaction)/Ecological Footprint

There are three objectives, with corresponding action plans:

1. To promote longer, healthier lives, with a target of ensuring an average life expectancy for a resident, wherever they live in the county borough, of at least the UK national average by 2030.

2. To promote fulfilled and satisfied lives, with a target of ensuring an average life satisfaction rating for a resident of the county borough of at least the UK national average by 2030.

3. To consume fewer resources, with a target of ensuring that the average ecological footprint for a resident of the county borough is 2.87 global hectares by 2030. (The current level is 4.81 global hectares per person – which is already the third lowest ecological footprint in the UK.)

Torfaen Council in Wales has now also introduced the same 'Living Better, Using Less' indicator, and in mid-2010 the nef was in discussion with Cardiff about doing the same.

Building on the Caerphilly indicator, the last thing the Camden Council all-party Sustainability Task Force did before it wound itself up in March 2010 was to suggest the creation of a combined indicator for judging the effect of Camden Council policies in terms of minimising the use of natural resources (ecological footprint), enhancing resilience (the ability of the community to withstand external shocks) and improving the well-being or happiness of residents.

BioRegional, the environmental organisation behind BedZED, the UK's largest low-carbon community, is working with the London Borough of Sutton to create a One Planet Living Index (see page 30) against which to judge the sustainability of council policies before they are enacted. The aim is to raise this yardstick to the level of the legal and financial sign-offs that all reports and policies are currently required to achieve.

And finally, Monteveglio, the town in Italy where the council and the local Transition Initiative wrote a joint Energy Descent Action Plan, has a wonderful goal of promoting a culture of simplicity and frugality to its residents!

The Local Wellbeing Project

The Local Wellbeing Project is a three-year initiative that started in 2007 to test out practical ways of improving public well-being through local authorities and their partners in three very different areas of the UK: Hertfordshire, Manchester and South Tyneside Councils. The final report, *The State of Happiness – Can public policy shape people's wellbeing and resilience?*, was published in early 2010.[6]

Against a background of intense pressures on public spending, the report recommends prioritising programmes that:

- Teach children resilience in schools – drawing on strong evidence that this improves academic performance and behaviour as well as employability of pupils.
- Promote opportunities for neighbours to get to know each other, based on clear evidence that this tends to enhance well-being.
- Provide support for isolated older people to help them create and maintain social networks, and reduce anxiety and depression.
- Shift transport and economic policies to encourage lower commuting times and allow people to spend more time with their families and friends.
- Reshape apprenticeships and other programmes for teenagers to strengthen psychological fitness to help young people find and keep work.
- Support families so parents are happier and children are less likely to face problems at home and at school.
- Promote activities that are simultaneously good for the environment and reducing CO_2, and make people feel better about their lives.

The authors themselves felt it was very much an interim statement because we're at such an early understanding of this crucial issue.

> "The Local Wellbeing Project has demonstrated how an emphasis on well-being in schools can be used as part of wider strategies to improve attainment and behaviour; how building social networks can increase well-being at the local level to underpin community engagement; how well-being can be used tactically to encourage pro-environmental behaviour; how a well-being focus can improve the impact of parenting

programmes; and how well-being can help older people live independently in the community for longer.

Thinking about well-being can shift the focus of policy. For example, well-being is often lowest in middle age, yet this is a time when people are least likely to use public services, being neither young adults nor children in education, neither parents with children, nor older people. Regeneration policies have often dispersed long-established communities, breaking down 'bonding' social capital and harming well-being even as the physical fabric has improved.

Our own work has encouraged us to want better measurements and understanding of the many dimensions of well-being. But it has also pushed us to want to develop a better understanding of the relationships between people's current well-being and their future prospects. For policy-makers it's vital to know not just how people are faring now, but also how prepared they are for the future. These questions of resilience and adaptability are particularly important during a period of recession and recovery and will be an important focus of the next phase of our work."

Local Wellbeing Project, 2010

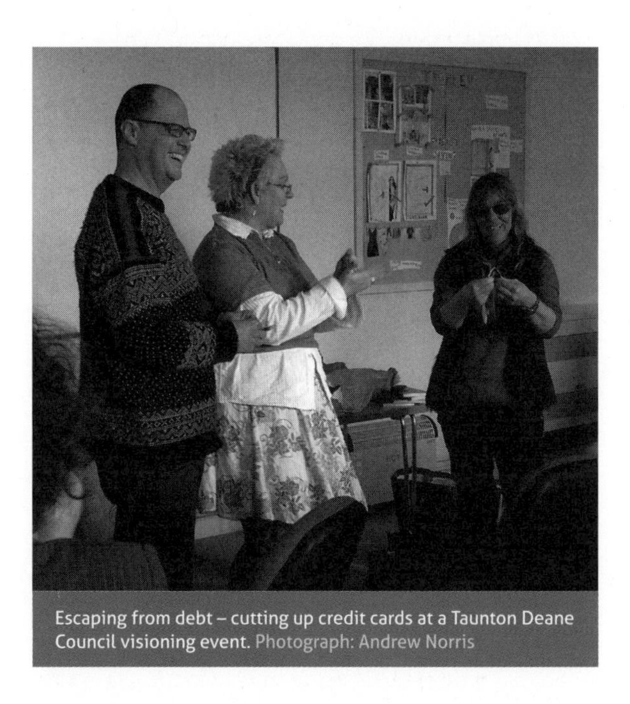

Escaping from debt – cutting up credit cards at a Taunton Deane Council visioning event. Photograph: Andrew Norris

» *Start a campaign to persuade your council's Sustainability Team to measure its policies and actions based on things that matter like happiness of residents or the ability of the community to withstand external shocks*

Chapter 14

GETTING ELECTED

Questions answered in this chapter:

- Do eco-activists need to get elected to move things forward faster?
- Do you have to sign up to conventional party politics?
- How exactly do you get elected?
- How do you influence an electoral campaign?
- How important is getting elected versus traditional eco-activism or Transition?

Do eco-activists need to get elected to move things forward faster?

The Transition movement has always had a healthy scepticism for traditional politics and political parties. In April, 2009 Ed Miliband, who had just been appointed Secretary of State for Energy and Climate Change, was invited to the annual Transition Network conference not as a speaker but as a Keynote Listener. In January 2010 the Transition Network released a 'Resilience Hustings' guide to engaging with election candidates[1] (see page 207). Again the focus was on keynote listening, on trying to help traditional politicians to understand the Transition way of thinking rather than making political demands or getting involved in the hurly burly of the conventional political process.

That's all good, but community groups can and should go further than that. Step Eight of Transition in *The Transition Handbook* is 'build a bridge to local government' because, as I have tried to set out in this book, there are lots of ways local government can help, even if only a few really have the bit between their teeth. And, like it or not, the current system is where we're at.

Only a handful (14 per cent) of those who replied to the online Transition Network survey created for the purposes of research for this book thought it was very important to get Transitioners elected, and most of them were already councillors. A further 35 per cent thought it was important to some extent. But only 23 per cent of those who replied said they would consider putting themselves up for election. Assistance with getting elected was right at the bottom of the list of things people thought were important – 35 per cent said it was not important and nearly half (49 per cent) said they needed no help on this score.

My view is that it is indisputably the case that you can get more done if you are inside local government as an

Quotes from those who replied to the Transition Network survey on local government

"Our local politicians are generally seen as corrupt and ineffective in the face of an old boys' network and old money. It is a difficult environment, and far healthier to subvert subtly than trying to co-opt. However, I support other candidates in other areas where the conditions are not so difficult."

"Councillors have a very wide range of responsibilities to their local communities, and they cannot devote themselves exclusively to Transition, as we can do – unless they were elected with Transition as the sole platform. Better to be a stakeholder bringing Transition to the Local Strategic Partnership and community meetings across the local area, and to be accepted as specialising in Transition."

"I have considered it, and may do one day, but as a Local Strategic Partnership member and chair of the local environment forum I already have influence and can achieve change. I would love it if we had some Transition councillors, but we should recognise that there are other routes 'in' to the process."

elected councillor, but it's certainly not a course for the fainthearted.

> "I think some people are going to be better at the getting elected thing than others. I really don't think every Tom, Dick and Harry can stand and get in and make a big difference. It needs a particular character type. You need to be quite patient and not get your knickers in a twist. You need to see where the chinks are, the ways in."
>
> Cllr Linda Hull, Independent councillor on Glastonbury Town Council and member of Transition Glastonbury

For me personally it was a tough four years. But I also know, and I think most people accept, that Camden Council has done more on this agenda as a result of my being elected and spending the last four years constantly pushing for action. You can argue about my methods, but you cannot, I think, argue that Camden would have done more if I had not been elected. And I'm sure the same is true of Paul Buchanan, Fi Macmillan and Linda Hull (see Chapter 3).

> "We should definitely be encouraging people involved in Transition groups to put themselves forward for election. Not only does it help inform a wider audience and promote understanding and support for the issues we are working on, it also builds bridges with other decision-makers,

It's great being a politician! Trying to persuade a group of environmental activists that some of them need to get themselves elected. Photograph: Mike Grenville

many of whom do want to support the principles we are working from but don't really know how to interpret these in local decisions."

Cllr Jacqi Hodgson, Green Party councillor on Totnes Town Council

"One of the things I have talked about a lot is the need to get involved. People complain about the make-up of the council, and I say, 'What do you expect?' You didn't vote, few people would stand, and then they moan about them. I regularly tell people they need to participate in the process, because we get what we deserve."

Mike Grenville, Transition Forest Row

This is not a Transition 'party line'. Nobody at the Transition Network level is planning to set up a Transition Party or recommend Transition candidates. Standing for election requires time, effort and emotional energy.

Formal power in a council goes: Leader (or elected Mayor), Deputy Leader, Executive or Cabinet Member, Chair of Committee (e.g. Scrutiny or Planning), Champion, backbencher in party of power, and lastly opposition backbencher, but the overwhelming bulk of the information and decision-making stops at the door of the Executive or Cabinet. To effect significant change you need to be in the ruling party or coalition, and ideally in the Executive or Cabinet. Backbenchers have very little power. Backbenchers in opposition parties have almost no power unless they have powerful friends in the community and/or the local media.

Of course informal power can take many forms, so at one level it is about how you go about persuading fellow insiders (councillors and officers) and outsiders (residents and journalists) to support your cause. On a good day everything clicks – you prepare officers, you flatter Executive egos, you get the press on your side, you organise deputations by and petitions of residents – you build up a multi-faceted campaign that enthuses everyone. On a bad day frustration kicks in because of the lack of information coming down from the Executive, the slowness and risk-averse nature of the bureaucracy and the apathy or lack of understanding of councillors.

Hopefully, if your group does decide to put someone forward for election, then you'll learn from the successes and failures of the eco-pioneers outlined in this book. But if you do go down this route be aware that your councillor candidates will need a lot of emotional support, because it's a gruelling business.

"I think it's ideal to have members of your Transition group on the local council. Several members of our Transition group go to the town council meetings and it's time well spent in terms of understanding of how the council operates and what their priorities are. Having a support committee for your Transition group within the council is also a good option, but these councillors have to be respected and influential members of the council. A close relationship with the Transition group and the town council is essential, and it's hard to imagine a group achieving long-term success without this."

Pat Bennett, Sustainable Somerton

Do you have to sign up to conventional party politics?

If you do decide to try to get elected should you stand as a representative of your local Transition or community group? Maybe, but I have to say that it's hard to get elected in local government if you're not in a political party. In some parts of the country there's a tradition of electing independents, especially at parish or town council level. But in most parts of the country, unless you are standing on a particular issue – like the closure of a local hospital or the need for a new school – something that has stirred the passions of a lot of local people – then you're going to need a political party. I'm not sure I believe in political parties as much as I once did, but they remain the principal vehicles for political power in this country.

Most environmentalists think about the Greens and the Liberal Democrats as holding views most similar to themselves. I know I did. I compared the two parties very carefully, decided they were very similar on environmental issues, but that I was likely to get more done in the Lib Dems. I was standing in Camden after all, where the Lib Dems were on the rise. Where the Greens hold the balance of power or are the largest opposition party, or where the Lib Dems are in power or are on the verge of taking power, it will make sense to join the Greens or the Lib Dems. But in 85 per cent of the town halls in the country it's the Conservatives and Labour who hold power.

So here's where I get a bit controversial. From the point of view of climate change, or human society's ability to cope with the end of cheap fossil fuels, it really doesn't matter which party you join so long as they have, or are likely to have, their hands on the levers of power. That may sound cynical and nakedly power-hungry, but it's more about the fact that we probably have, if we're lucky, just five to ten years to make a difference on climate change and to prepare for peak oil. With apologies to the opposition parties in all these areas, if you live in Wandsworth or Kent you should probably join the Conservatives; in Manchester or Newham you might want to join the Labour Party; in Kingston-upon-Thames or Newcastle you should think about joining the Lib Dems; and in Brighton or Lancaster you could join the Greens.

There is an additional benefit – the mainstream parties need changing from the inside. So my advice is find out who's in power or on the rise in your area and look at the recent election results on Wikipedia to check trends. It's OK to join a party that looks like it's about to arrive in power. Just do not join one that cannot possibly be in power, at least not if you want to make a difference on climate change and peak oil. We simply don't have time.

Join the party in question nationally and then make contact with the local party. They'll be delighted to see you. Show willing and you'll soon be sounded out as a possible councillor candidate. It's shockingly easy to become a councillor candidate. The job is so all-consuming and so poorly remunerated that very few people want to do it or feel able to spare the time to do it. It also earns few thanks from residents who tend to tar all politicians with the same brush.

Of course, from the point of view of a healthy democracy it's terrible that it's not difficult to become a councillor candidate, but it does make the task of getting people who care about the planet into councils across the country a bit easier. As I said earlier in this book, I went into politics because of my concern about climate change. I later understood the significance of peak oil and that strengthened my desire to effect change.

Once you're in the party, start pushing on environmental issues – although you might want to wait until you've been chosen as a candidate before you push all of the policies in this book! Mind you, things are moving so fast that things considered radical in Camden in 2006 may soon be completely mainstream. I was considered to be quite mad when I started

We are the world – talking to primary school children in Camden about what we need to do to ensure that they have a planet to live on.

talking about the need to eat less meat and better-quality meat. Now people such as Lord Adair Turner, Chair of both the Financial Services Authority and the Climate Change Committee, say the same thing.

I used to give presentations in schools and be greeted by blank faces when I said they should be growing food, cycling to school or recycling their rainwater. Now they regularly lead me off to the garden to see their carrots, their rescued battery chickens or their real-time energy monitors. When I started talking about councils lending money to residents to pay for energy efficiency improvements and recouping it from energy bill savings my officers said it was far too risky. Now it's being piloted across the country. When I said all buildings should be built or refurbished to the Passivhaus standard, the only proven energy efficiency and comfort standard in existence, officers said I was being too prescriptive. Then the Building Research Establishment wrote a Passivhaus primer that said there was no other way to reach Level 6 of the Code for Sustainable Homes.

Things change fast. Yesterday's radicalism is today's mainstream is tomorrow's old hat. My advice to you is: be sure of your argument. Look for examples elsewhere. Make sure nobody knows more than you do or that you can call on persuasive experts to support your case. Above all – never give up.

Transition and the Greens

There's a slight risk that some political parties, particularly the Greens, will become over-associated with the Transition movement. In a sense that's quite understandable. Ever since the Greens' leader Caroline Lucas wrote a paper on peak oil in 2007 the party has been closest in its thinking to the aims and ambitions of Transition. However, there are towns in the north-west of England where the ruling Labour elites have sidelined Transition by claiming that it's simply a Green Party front.

> "I do think there are risks for the Transition movement more widely because an overly close association with the Greens can be alienating for everyone else. I think the Transition movement should stay totally apolitical."
>
> Fi Macmillan, Green Party Councillor, Stroud District Council and member of Transition Stroud

I almost never say which political party I'm from – I don't think it's relevant. Politicians who are prepared to work with other parties and with environmental community groups should be embraced, but Transition groups should all be wary of the traditional party-political response, which is to pick the best headlines to use as political differentiators and worry less about the follow-up action.

Transition Kentish Town in Camden started up in January 2010 and the initiating group quickly found itself trying to answer the question of whether an enthusiastic local councillor should be allowed to join the group. There are a number of councillors coming through who are primarily motivated by climate change, peak oil, etc., such as the ones in Chapter 3 of this book. I think the key thing is motivation. Why are the councillors in question taking part? If you suspect their motives – i.e. they're primarily doing it to reap electoral benefits – then it's best to keep them off the steering group and instead welcome them with gusto to all events. They can be useful – no question. I learnt a lot in four years at Camden; I now know exactly who to ask for anything and the limits of what's possible. So, keep an open mind, scrutinise motivation carefully, engage in honest conversations, but whatever you do, don't alienate them.

How to deal with the BNP

The British National Party has also been sniffing around Transition, which presents a different challenge. The BNP seems to be trying to take ownership of the peak oil debate. For them any crisis is an opportunity to further their divisive ideology. The BNP creates and works on people's fear. It's important that the people who want to see a good, positive and inclusive Transition are able to counter the lies and hatred of the BNP and put forward that positive vision. Thankfully, despite the fact that immigration was centre stage in the 2010 local and general elections, the BNP didn't make any headway, so maybe, hopefully, this problem is receding.

How exactly do you get elected?

Getting elected is a slog but it really isn't hard. If you and your fellow candidates can knock on enough doors, send a letter to everyone you meet explaining what your position is on whatever that voter cares about, send a letter to every street saying what you would do for them, and collect emails and telephone numbers as you go – then frankly it's easy. If you're challenging an incumbent in another party, then the more you can do without being seen the better. Flooding letterboxes with generic leaflets is old school. They usually go straight in the recycling bin.
You need to knock on doors and start a dialogue, which you can continue by post, email or telephone. There's still a certain amount of tribalism around for the two main parties, but if voters feel that you actually care about their issues, then they will vote for you. Identification of issues, not voting intentions, is the key.

Lots of demographic information can be gleaned on the doorstep. A pram in the hallway means young children, which means potential interest in education policies. A frail voice over the intercom means someone who'll probably be interested in issues affecting older people.

Right up until polling day in May 2006 nobody outside the safe Conservative ward of Belsize thought the Lib Dems could win the ward, and that included the leader of the Camden Lib Dems. Indeed, in the days running up to the poll several members of the Lib Dems said they thought it was indulgent to let me try. We were in a distant third place in 2002 with an average vote for the three Lib Dem candidates of 440, far behind the Conservatives (average vote 1,021) and Labour (average vote 726). We had no ward party to speak of. We had no candidates apart from me until three months before the election. Our third candidate only joined the Lib Dems on the final day of registration. We should never have won.

But we did win. Laura and I roped in 50 friends and family to help out, none of them Lib Dem members. We knocked on 2,000 doors. We asked people what they were concerned about and then wrote to every one of those 2,000 people to explain our policy on their particular issue. We wrote targeted street letters about potholes and street lighting. We targeted the Labour and Green vote ruthlessly. We used issues like the threat of the local police station being closed and draconian traffic wardens to win the soft Tory vote. And the Tories didn't see us coming until it was too late. Maybe we were lucky. But I don't think so. Energy and enthusiasm are what win local elections.

Of course it should be pretty easy to convince voters that you rather than Candidate Bloggs is credible on the environment. After all, not only will you know what you're talking about on that score, but you'll also be able to point to the way you live your own life. Voters are turned off by hypocrites. Far too many politicians do not walk the talk. You can be different.

It goes without saying that you need to have answers to a few things other than climate change and peak oil! But your local party can supply you with those. Or maybe you can come up with better lines and persuade the local party to use them.

It helps if you can create a relationship with the local press. This is hard for a newcomer, so at the very least the rule should be to write letters to the local papers as often as possible. What I try to do is recycle my thoughts between the local papers, my blog, personal letters and emails, street letters and email newsletters to constituents. By the time I left office in May 2010 I'd built up an email list of 1,500 voters. If not for the churn rate of voters in Belsize (20 per cent of the electorate changing every year!) my email list would have been much bigger.

Email newsletters that are about local news rather than party politics work well. If you build up a reputation for keeping residents informed about events in the area, then they'll accept the occasional plea for political support. Barack-Obama-style campaigning techniques, where you ask librarians to contact librarians or union members to contact union members, work much better than getting somebody from a conventional political party to ring up a voter.

Keep an eye out for campaigns that can put you in touch with more than one elector. Planning applications sometimes provide a good opportunity to lead a campaign on behalf of residents. Street petitions are an excellent way of gathering contact details for potential supporters – but people will only sign them if the petition is actually about something meaningful such as the threatened closure of a post office or a hospital.

How do you influence an electoral campaign?

In the run-up to the 2010 general and local elections the Transition Network produced a guide to running Resilience Hustings, an attempt to help parliamentary candidates to understand the Transition movement[2] (see box, opposite). The guide explained how to run

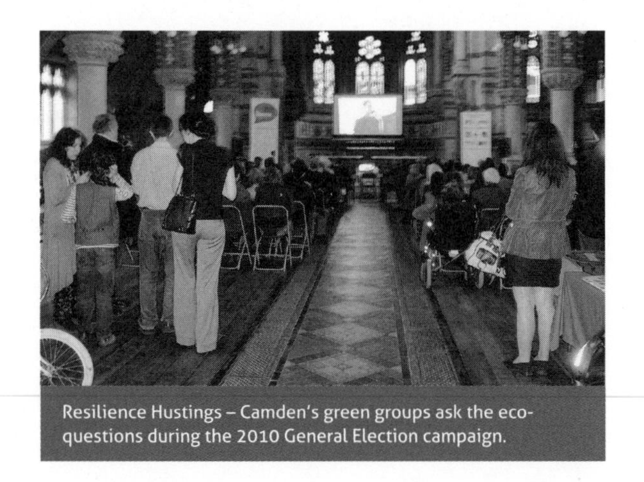

Resilience Hustings – Camden's green groups ask the eco-questions during the 2010 General Election campaign.

How Resilient Are Our Political Candidates? Embedding Transition in the Political Process A Hustings Primer from the Transition Network 2010

Can your Transition Initiative help shape politics as we head towards the 2010 elections?

Should Transition Initiatives get involved with formal politics? Some decry existing political structures, preferring to get on with building a positive future; others call for Transitioners to get themselves elected to help make change happen now. As with much of Transition, there is no right way, only the way that you choose – but if you do want to get involved in the local and general election campaigns in 2010, then here are a few suggestions. You could make quite a difference to how electors and elected see the world and their roles in it . . .

What is a husting – and more importantly, what makes a good one?

The dictionary definition is "a platform on which candidates for Parliament formerly stood to address the electors". In essence, it is an event, attended by the political candidates for a constituency, at which the public are able to ask them questions.

At a good husting:

- the speakers and the audience have to have clarity about the ground rules and what everyone is aiming to get out of the event
- the Chair has to play an active, facilitative role, seeking to make sure no one dominates, as many people as possible are heard, and people answer questions clearly and concisely
- it will be energising – after all, it should attract engaged people, thinking and communicating about crucial issues.

Here are some suggestions for making your event as successful as possible:

1. Invite your local parliamentary and council candidates to Transition events as Keynote Listeners. At the end, ask them to speak for five minutes about what they have learnt and about how their personal or party platforms relate to the themes of Transition. Then organise a husting in the run-up to the election, either parliamentary or ward-based, depending on the size of your Transition Initiative and your ambition. The aim is to ask each candidate to explain and debate their approach to developing resilient local communities/economies prepared for climate change and peak oil. This booklet is designed to prepare you for that husting.
2. You might start by showing some or all of the film *In Transition*. If you do, then give the candidates front-row armchairs and make them feel at home. After the film ask each candidate to reflect on how their policies relate to the film's themes.
3. Keep a record of the questions and answers at the final candidate husting as well as the candidates' names and political parties.
4. Ask someone from your group to be a contact point and feed the results back to the Transition Network so we can get a sense of what's happening across the country.

It's important to recognise our potential role in shaping the broader debate on issues that transcend normal party politics, and your husting event will be a key contribution to that.

See here for the rest of the document, which defines resilience and gives answers to the Nine Buts to Transition:

http://transitionculture.org/wp-content/uploads/Final-designed-hustings-version-2.0.pdf

Transition Town Totnes did their Resilience Hustings slightly differently: they did it as a World Café. All the candidates were given their own table and voters circulated asking questions, joining in discussions or simply listening. Several of the candidates afterwards said it was the most challenging aspect of the entire election campaign!

hustings, defined resilience, and included nine Frequently Asked Questions that might arise at such an event – the Nine Buts to Transition.

How important is getting elected versus traditional eco-activism or Transition?

Some are worried that Transition Initiatives are sucking away activists from traditional environmental pressure groups such as Friends of the Earth.

> "I am concerned that Transition Bath, with its focus on local, has taken new recruits from our local FoE group. I think there is a lot of respect from councils and officers for FoE nationally and there is a lot of resources for working through the planning system, working with local councils, sustainable communities, etc. If Transition continues to promote itself politically, and continues to have such positive initial enthusiasm from the public, local groups of FoE will start to fold. Undermining (however unintentionally) the superb network of local groups and campaigning power that FoE has could be quite detrimental to your goals and theirs."
>
> Scott Morrison, Friends of the Earth Bath

My sense has always been that there's a bit of both in Transition – new people and more seasoned activists (see box opposite for evidence of this).

But Transition's in a different place from Friends of the Earth. In my opinion both are critical. The Transition agenda is primarily about turning a positive vision of the future into reality. It's about getting on with doing things that recreate community and trust. Physical manifestations of the project are key.

Friends of the Earth (FOE) is a magnificent campaigning organisation. Its 'Big Ask' campaign won us the 2009 Climate Bill, which requires UK governments to set and meet CO_2 emissions targets. FOE's lobbying of traditional politicians on environmental issues is probably second to none. But this isn't an either/or situation.

From 2006 to 2010, while I was an elected member of Camden Council, I was also an active member of Friends of the Earth and Transition Belsize, and a paid-up supporter of Greenpeace and the Soil Association. I also participated in Climate Camp every year from Heathrow onwards and I regularly wrote to my MP and to the newspapers. And since July 2009 I've worked on the 10:10 Campaign, coordinating council sign-ups. These things are not exclusive – they're all linked. But they're all needed. You can be part of all of them or just concentrate on one, like being part of the Transition movement. The main thing, as far as I'm concerned, is that you do something, and that together we are stronger than we are alone.

Testing the candidates – a World Café in Totnes during the 2010 general election campaign. Photograph: Rob Hopkins

Activism profile of Transition Norwich members

A 2009 study of Transition Norwich members by Gill Seyfang of the School of Environmental Sciences at the University of East Anglia[3] revealed that for nearly a third of the members (32 per cent) Transition Norwich was the first local environmental group they had been involved with. A further 19 per cent had previously been involved in similar activities, but were not involved with any other groups at the time of the survey. This finding is significant as it demonstrates that the Transition movement is capable of enrolling and engaging new people (or re-engaging ex-activists) in local environmental groups, rather than simply re-badging existing campaigns and activists (although it clearly does this as well).

A third (33 per cent) were currently involved with other local environmental groups, and 16 per cent were also involved in groups that were quite different from Transition Norwich. These included local permaculture groups, slow-food groups, local organic gardening groups, Freecycle, farmers' markets, the Green Party, etc.; and also included organisations such as RSPB, Friends of the Earth, the Quakers, the Ramblers, Greenpeace, Rising Tide and so on, which have quite distinct goals and modus operandi from the Transition movement.

The survey asked members what they thought differentiated Transition Norwich from other local environmental groups, and the responses indicated that there were three important distinguishing features. The most frequently reported of these (cited by 50 per cent of respondents) was the interlinking of a broad range of issues under a single 'umbrella' brand, for example: "I don't see it so much as a new group but as a process that can bring together a lot of pre-existing activity, catalyse new work, and create new interest in a range of seemingly disparate issues." Another said: "it is a wide umbrella group of diverse interests, but with a common goal to act locally . . .", and that "it is comprehensive in that it covers all the different themes in a separate but interlinking way" and another said "it has a wider vision, brings together many different strands".

Second (mentioned by 29 per cent) was the focus on positive practical action as opposed to negative-seeming protesting or campaigning, as explained by this respondent: "It's not about creating a 'them' and 'us' opposition, though it can and does challenge existing orthodoxies. Its primary means of motivation is offering a positive vision that inspires people to join in, rather than inviting people to join in with demonising and scapegoating a group or institution. 'What are we for?' is a much richer and empowering position than 'Who are we against?'." Another explained that it was about "doing things positively to make a difference, rather than campaigning", and another said "it's working for, rather than against, something".

The third most-cited difference (21 per cent) was the grassroots nature of the organisation, with its emphasis on community empowerment, such as "It focuses on doing it for ourselves, rather than persuading others", "facilitates bottom-up change that does not rely on government", "helping empower people to change and be catalysts for change" and "TN is grassroots, therefore less exclusive or prescriptive".

Further features mentioned included the focus on local solutions (10 per cent), the movement's mass appeal (10 per cent), the fact that the local group is part of a wider movement (8 per cent), the community-building aspects of the group (6 per cent), and the element of inner change that Transition highlights (2 per cent). It is clear then, that Transition Norwich is seen as being different from other local groups, with specific characteristics (not single-issue, a focus on positive action, and bottom-up community empowerment) that make it complementary to existing initiatives, and an attractive proposition.

Actually I would go much further than that and say that I would like to see Transition reaching into everybody's life whether or not they engage in environmental activism. We need to meet people where they are, not where we'd like them to be. We need to help them to act according to Transition principles in their workplaces, in their communities and in their homes.

So I would say that to push this agenda forward truly effectively we need four forces working together:

- Eco-activists working from the inside of a council, e.g. elected members or officers.
- Visible manifestations of a positive future without cheap oil, by Transition groups or people acting according to Transition principles in their workplace, community or at home.
- Political lobbying, e.g. Campaign against Climate Change, Friends of the Earth, Greenpeace.
- Direct action, e.g. Climate Camp, Plane Stupid, Greenpeace.

Onwards and upwards – Rob Hopkins addressing the annual Transition Conference in June 2010

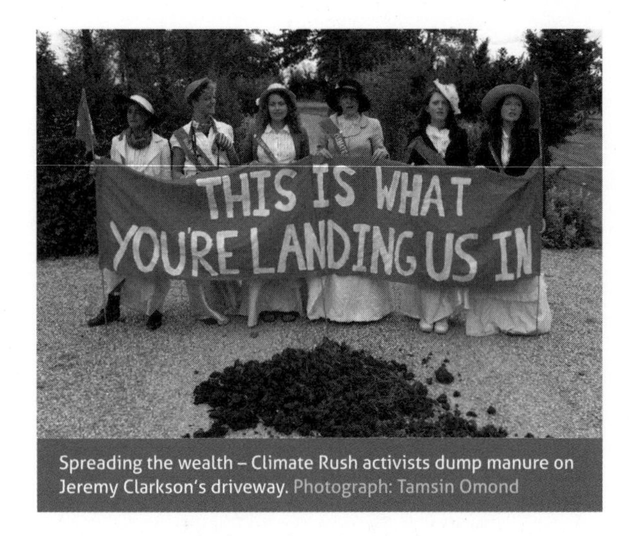

Spreading the wealth – Climate Rush activists dump manure on Jeremy Clarkson's driveway. Photograph: Tamsin Omond

If I were trying to win a million-pound business deal, then I'd be pushing on every possible lever: I'd be making sure I was listening to the concerns of every significant player or group in the organisation and trying to offer a solution to all of them. If I were trying to lay my hands on a good story as a journalist, then I'd be trying every possible source. We need to do the same in environmental politics. We need to push from the outside, the inside, the left side, the right side, the top side and the bottom.

We cannot wait. We need to act now. All of us. And communities and councils, acting together, can do more than you think, as I hope this book illustrates. So let's get on with the task in hand and create a positive, low-carbon future, one where we learn to live in tune with the natural world rather than in opposition to it. If governments can't, won't or don't get their acts together, then we can and must show them how.

On 3 December 2009 I received a call from Pete Cooper of Congleton Town Council in Cheshire. He'd just signed the council up to the 10:10 Campaign. I congratulated him and asked what area of their emissions they'd be trying to reduce. He replied that the main energy user was the town hall, which they'd just had energy-audited, and said they'd be trying to reduce that by 10 per cent in 2010. But what really moved me was his next sentence. He said he felt it was important for the Town Council to lead on this, to try to inspire others, because they were planning to become a Transition Town and he felt this was in keeping with that goal.

If he'd been able to see my response at the other end of the telephone – an ear-to-ear Cheshire Cat grin – he'd have thought I was quite mad. But his words felt like the ultimate proof that, working together, communities and councils can make a difference.

In the words of the anthropologist Margaret Mead:

> "Never doubt that a small group of thoughtful, committed citizens can change the world. Indeed, it is the only thing that ever has."

I think that's my favourite quotation of all time.

United we stand – the launch of the 10:10 Campaign on 1 September 2010 at the Tate Modern in London. Photograph: 10:10 Campaign

Appendix

EXECUTIVE SUMMARY OF THE TRANSITION AUDIT OF SOMERSET COUNTY COUNCIL

Conducted by Daniel Hurring (2009), Operations Director, Natural Communities Community Interest Company

1) Context of the Report

This Transition Audit was commissioned as a result of Somerset County Council's (SCC) historic decision to become the UK's first Transition Authority. In July 2008, at a meeting of the full council, it was unanimously decided to support a resolution confirming this move and to review services and budgets to make it happen. This commitment emerged because of the growth of the Transition movement in Somerset in particular, but across the UK as a whole.

A report was commissioned by the then Leader of the Council through the Sustainable Development Team to establish the extent to which the Council's current strategies and policies were consistent with Transition objectives in 12 theme areas.

In order to produce the report an initial survey was circulated to selected respondents within SCC. These were chosen from the existing knowledge of officers and from research carried out by members of the Sustainable Development Team. This process was as comprehensive as possible in the time available.

The main aims of this report then are to:

Audit the existing services of the Council in light of the commitments made in the Transition Resolution and the objectives of the Transition movement itself

Establish the next steps Somerset County Council are to take in their becoming a Transition Authority

Recommend further actions and initiatives, both general and category-specific, that would assist the Council in becoming a Transition Authority

The Audit is intended as a guide rather than a set of hard and fast rules as to what it is to be a Transition Authority. It does not try to be prescriptive, but instead offer options on how the resolution could be taken forward, with the emphasis for leadership placed upon the County Council itself rather than the Transition movement

2) The Strategic Context

The Transition Audit is written within the context of the Sustainable Community Strategy (SCS). This document, produced by the Somerset Strategic Partnership, outlines the key strategic objectives for Somerset over the period until 2026. The SCS uses the theme of 'Imagine Somerset in 2026' a technique that is shared with the Transition Towns movement. It includes a series of themes and 'Big Ideas' many of which also merge well with Transition, including establishing Somerset as a 'Green County', Living Sustainably and Making a Positive Contribution. The alignments between the SCS and the Transition objectives mean that the former naturally provides a backdrop to this report. The aim of this report, in turn, is to complement rather than compete with the SCS.

3) The Need for Change

The report looks at the wider context in which this Audit takes place, in particular at the twin threats of Climate Change and Peak Oil, which are the main drivers behind the Transition Towns movement. The former of these concepts, Climate Change, already plays a major role in policymaking at all levels, and has fuelled the idea of a 'low-carbon economy', which is expounded in such national strategy as the Department for Energy and Climate Change's 'Low Carbon Transition Plan'.

Peak Oil is a much less researched phenomenon that is only now coming to the attention of governments, though many scientists have warned of it for years. Peak Oil makes inevitable what Climate Change makes essential. Peak Oil theory brings into focus our dependency on fossil fuels, both across the County and on a much wider scale. It highlights that on all levels, changes of habit are rapidly needed so that when the inevitable happens, and fossil fuels become prohibitively expensive, we are prepared. This report highlights many ways in which the County Council needs to change its own habits, where it is already doing so, and offers a few potential solutions.

Transition's other major motivation, perhaps even beyond these two factors, is the need for increased local and community resilience. This means securing our food supplies and our energy sources, building local low-carbon transport networks, refocusing our economies and building a sense of shared purpose and a shared vision of the future. As the Local Authority responsible on so many levels for the welfare of those under its duty of care, Somerset County Council needs to play its part in building this. Again, this report highlights many ways it is already doing this, and many ways it can continue to do so.

4) The 12 Theme Categories

The Audit uses 12 themes across which the County Council is examined. The first seven of these are based on *The Transition Handbook*, by Rob Hopkins, the other five upon areas of responsibility for the County Council. A summary of findings in each of the categories is contained below.

4.1) Food & Farming: The Audit highlights the value of assets already in hand, in this case the County Farm Estate. It recommends the implementation of a local food policy within the Estate, and an increased focus on environmentally-friendly and organic farming. It also identifies the opportunity for the establishment of a working 'model' Transition farm project as part of the County Estate, highlighting low-carbon farming practices.

The report also commends the Council's previous positive investment in local foods and its continued support of the Farming and Wildlife Advisory Group. It highlights the importance of supporting local food networks and initiatives.

4.2) Health and Well-Being: The report recognises that Health is to a large extent a non- statutory responsibility for the County Council. Nonetheless, it identifies the need to join up policy more effectively, encouraging active lifestyles through engagement with community activities – such as creating a community garden, tree planting and developing play facilities. Focus is seen to be needed on these forms of activities being established in urban areas especially, where deprivation and poor health are often at their highest. The report also recognises the opportunity for basic re-skilling of populations, in skills that also benefit health, such as home cooking for nutrition, and caring for a garden.

4.3) Education: The report highlights the positive steps made by SCC in raising awareness of Climate Change, but recommends the same is done for Peak Oil and our fossil-fuel dependency. Beyond this, it sees the priority as being the teaching of skills and knowledge that allow people to become useful members of a sustainable community. It recognises a need for more coherency in the way sustainability is taught and practised in schools, with an increase in out-of-classroom activities in particular. Finally, it calls for the ongoing University of Somerset programme to prioritise 'green' principles and courses.

4.4) Economy: Building strong local low-carbon economies is the recommendation of the report, with businesses and industries rooted in the County and in the South West. It highlights the opportunities inherent in the current reform of society caused by the imminence of Climate Change, not least in the development of green technology industries and the localisation of resource networks. There is a need to review economic policy in light of peak oil, where a less energy-rich future may be on the horizon and where continual growth may not be the likely outcome. In terms of the internal workings of the county council, reviews of what areas of its services could be delivered differently to strengthen the local economy are necessary.

4.5) Transport: The report recognises Somerset's forward thinking Local Transport Plan, supported by documents such as the Cycling Strategy and initiatives such as Moving Somerset Forward. However, it also points out the need for more radical changes in this sector. Some of the proposals include reviewing the weakness of the local rail system, the opportunities presented by new technologies such as electric vehicles, and the desirability of a revived waterways system. The greatest shift in transport use, though, is predicted as being the switch to a more communal mind set in private vehicle use.

4.6) Energy: The report reviews the ongoing work of SCC and, in particular, the Sustainable Development Team in this area. It acknowledges the large number of positives, including a renewable energy installation on the County estate – especially in schools, energy-from-waste initiatives and internal awareness-raising programmes, to name but a few. Improvements are seen to be most important in helping support and develop community energy schemes as a complement to the larger projects. The report acknowledges that this is made more difficult by target-focused delivery, which prioritises big projects, but also promotes the need for a balanced approach. The other improvement necessary is in better communication of the work being done.

4.7) Housing and Development: This area is viewed as having a particular need for partnership working with District Councils, in implementing sustainable planning regimes. Sustainable construction and development is largely overlooked by the SCS, which is noted, and there is an urgent need for SCC to make a stronger case for standards in this respect, especially in choosing project partners. SCC's primary role is seen as being to set strong guidelines and parameters that raise the bar in the County. More direct action can be taken through supporting Community Land Trusts and promoting sustainable construction through conferences and educational events.

4.8) Building Community & Local Governance: Attempts are needed to address the rising disillusionment with local democracy and fragmentation in local communities. The report recommends utilising the relationship with Transition Towns to develop forums for discussion in communities and to establish local organisations that bring cohesion – such as community development trusts, land trusts, energy partnerships and growing schemes. Ensure council members are trained in the objectives of the Transition Movement and Resolution as part of their role as Community Champions. Finally, local events are identified as key factors in bringing communities together.

4.9) Social Cohesion: The report identifies the tendency to firefight problems in this sphere – treating a sick society rather than creating a healthy one. It looks at core issues including elderly and youth disaffection – solutions for the latter being suggested as increased emphasis on youth involvement in local government through forums and engagement projects, and improving their image in the media and Council documents. There is also a need to impart improved training to members in equality and diversity policies. Community projects, such as Transition-related ones, are an opportunity for disaffected people to become involved in meaningful activities which can build valuable skills and serve as a base for promoting good mental and physical health.

4.10) Waste and Water: The SCC approach to waste is exemplary. Where improvements can be introduced most effectively is in the farming sector, which needs a sustained campaign of awareness-raising and investment. The recommendation of the report is not to allow large-scale waste developments to overshadow smaller community efforts – for instance in energy-from-waste schemes. Technologies related to waste water are highlighted as important, especially in new development where effective natural methods can be used. Finally, the unique End Uses Register (for products disposed of in recycling centres across the county) available locally, which is innovative in itself, could be improved to follow the chain of waste to its final destination as a reused material/product.

4.11) Land and Environment: SCC has a strong commitment to the Natural Environment, expressed in its developing set of strategies under this heading. Many of these are in consultation stage, and there is still the opportunity to embed Transition principles in them, to further enhance their value. In this topic area, it is more a matter of minor alterations than large- scale change of policy that is needed. It is recommended to ensure that all tree and woodland planting is focused around native and productive trees – be these for food, fuel or as a building material; the strategy is also viewed as an opportunity to explore multi-use community woodland as a solution to a variety of issues. Again, the County Farm estate is highlighted as a convenient resource for demonstrating principles of environmental management of land for diverse use.

4.12) Tourism and Culture: Key areas identified in this theme include the opportunities presented by the Cultural Olympiad to highlight the 'green' ethics of Somerset; to look to support a major green tourist attraction in the county – perhaps a rejuvenated Waterlinks project; increase ways of supporting local community events and creating local Tourism partnerships that engage residents and allow for increased community ownership and design. Heritage, particularly as regards traditional skills and crafts, is also highlighted as an area where Transition principles find comparison.

5) Further Actions and General Recommendations

5.1) Next Steps: An action plan, perhaps implemented over 3 years, is recommended to move forward the bid for Transition Authority status. A time limit on activities is essential in the light of the impending dual threats of climate change and peak oil. First year outcomes should be agreed upon as soon as possible, with annual actions and milestones to follow. The action plan should be formed internally, though with reference to this Audit and in consultation, on some level, with Transition Somerset. The other step, highly recommended to be initiated in the short term, is the employment of a Transition Officer to liaise with local groups and ensure the action plan is successfully implemented.

5.2) Planning for Energy Descent: A longer-term strategy is needed to ensure that energy descent is successfully planned for. A working group should be established to begin compilation of an Energy Descent Action Plan (EDAP). Simultaneously, the work being carried out on the Resolution should be pushed to the forefront of policy and procedure, if only to ensure that the rapidity of changes required is understood. Further training of members and officers in the principles of Transition should be organised and offered, perhaps even made mandatory to ensure mutual cross-departmental understanding. Finally, a review of working practices is required to eliminate areas of conflict with principles of sustainability and Transition, and to ensure that best practice is achieved across the board.

5.3) Seeking Partnerships: SCC has the option to liaise effectively with District and Borough Councils to attempt to achieve cross-tier partnership working on achieving Transition goals. Areas where this is particularly necessary include Planning and Land Use. Other partnerships on the Transition Resolution should be sought with all bodies involved in the Somerset Strategic Partnership in key sectors including Health, Environment and Economy. Externally to the County, it is recommended that an Annual Conference (possibly via video) is held between all councils in the process of Transitioning and all those interested in doing so. SCC could take the lead on this, thus exporting their model and raising their profile.

5.4) Spreading the Word: One of the primary benefits of the Transition Resolution is the opportunity to set an example to private business, public bodies and individuals. SCC's adoption of the Transition Principles sends out a positive message and allows others to follow their lead. Simultaneously, with SCC there is the need and chance to spread the word amongst its own departments, many of which will not naturally gravitate towards sustainability and local resilience and have no reason to do so.

5.5) Community Engagement and Ownership: Throughout the report, there is a recurring theme of community engagement and ownership. Transition's popularity is partially born of the desire of local populations to have some measure of control over their immediate environment. Local governance does not always succeed in this and many people do not feel represented by the standard electoral system, nor by the private interests of large commercial bodies such as property developers. Accountability and the development of community-owned schemes – including energy, land trusts & agriculture – will provide a greater sense of democracy and produce a more effective Transition.

5.6) Working with the Transition Movement: Members, and the Transition Resolution, would be best served by providing guidance on working with Transition groups as part of a wider training package. Reports of surveyed officers continually identified this during the process of producing this document. There is also great benefit to the Council in engaging further with Transition Somerset in the development of tools and practices that will better enhance the Transition Authority status and provide more effective means of community engagement and achieving local resilience in light of Climate Change and Peak Oil.

5.7) Principles of a Transition Authority: Transition Authority status is not defined, but some general principles are needed. These can be found in Section C1.1. The principles are based on those of the Transition network as a whole and incorporate awareness raising (of Climate Change and Peak Oil), developing local resilience, easing the way for Transition groups to flourish and aiding them in their progress, assessing council services against Transition principles, changing provision found to be in conflict with the status (where this is compatible with the other requirements of a council and not in conflict with statutory duties) and a commitment to self-regulate and self-assess progress, whilst reporting this progress to the Transition Network.

REFERENCES

Foreword

1 Shuman, M. (2000). *Going Local: creating self-reliant communities in a global age.* New York, Routledge.

Introduction

1 Heinberg, R. (2007). 'Bridging peak oil and climate change activism'. *Energy Bulletin*, Jan 2009. www.energybulletin.net/node/24529.

2 Crompton, T. & Kasser, T. (2009). *Meeting Environmental Challenges: The role of human identity.* WWF-UK: Godalming, Surrey.

3 Hopkins, R. (2008). *The Transition Handbook.* Green Books: Dartington.

4 Travers, T. (2009). 'Cutting Carbon Locally – and How to Pay for It'. Friends of the Earth: London. http://www.foe.co.uk/resource/briefings/get_serious_finance_report.pdf.

5 Osborn, S. et al. (2009). 'Building a Positive Future for Bristol after Peak Oil'. www.bristol.gov.uk/ccm/content/Environment-Planning/sustainability/file-storage-items/peak-oil-report.en.

6 Oil Depletion Analysis Centre (ODAC) / Post Carbon Institute (2008). 'Preparing for Peak Oil: Local Authorities and the Energy Crisis'. www.odac-info.org/sites/default/files/Preparing_for_Peak_Oil_0.pdf.

7 Lerch, D. (2008). *Post Carbon Cities: Planning for energy and climate uncertainty.* Post Carbon Press: Santa Rosa, California.

8 'Estate agents and politicians among least trusted professions'. *The Telegraph* online, 1 April 2009. www.telegraph.co.uk/news/newstopics/howaboutthat/5085369/Estate-agents-and-politicians-among-least-trusted-professions.html.

9 King, Rev. M. L. (1967). 'Beyond Vietnam: A time to break silence'. Speech delivered by Dr Martin Luther King, Jr., on 4 April 1967, at a meeting of Clergy and Laity Concerned at Riverside Church in New York City. www.hartford-hwp.com/archives/45a/058.html.

Chapter 1: A vision of a Transition council

1 Hansen, J. (2009). *Storms of my Grandchildren: The truth about the coming climate catastrophe and our last chance to save humanity.* Bloomsbury: London.

2 Ibid.

3 Heinberg, R. (2003). *The Party's Over: Oil, war and the fate of industrial societies.* Clairview Books: Forest Row, East Sussex.

4 Osborn, S. et al. (2009). 'Building a positive future for Bristol after Peak Oil'. www.bristol.gov.uk/ccm/content/Environment-Planning/sustainability/file-storage-items/peak-oil-report.en.

5 Oil Depletion Analysis Centre (ODAC) / Post Carbon Institute (2008). 'Preparing for Peak Oil: Local Authorities and the Energy Crisis'. www.odac-info.org/sites/default/files/Preparing_for_Peak_Oil_0.pdf.

6 City of Portland Peak Oil Task Force (2007). 'Descending the Oil Peak: Navigating the Transition from Oil and Natural Gas'. www.portlandonline.com/bps/index.cfm?a=145732&c=42894.

7 LGA (2008). 'Volatile Times: Transport, climate change and the price of oil'. www.lga.gov.uk/lga/aio/1335142.

8 IDeA website. 'Low carbon futures'. www.idea.gov.uk/idk/core/page.do?pageId=19882346.

9 Wikipedia. *Open Space Technology.* http://en.wikipedia.org/wiki/Open_Space_Technology.

10 Layard, R. (2005). *Happiness.* Penguin: London.

11 Brangwyn, B. & Hopkins, R. *Transition Initiatives Primer.* www.transitionnetwork.org/resources/transition-primer.

12 BioRegional/WWF (2008). *One Planet Living.* BioRegional/WWF: Wallington, Surrey.

13 Monbiot, G. (2009). *Death Denial.* www.monbiot.com/archives/2009/11/02/death-denial.

14 Worden, W. (1983). *Grief Counselling and Grief Therapy: A handbook for the mental health practitioner.* Tavistock Publications: London.

15 Kübler-Ross, E. (1969). *On Death and Dying.* Touchstone: New York

16 Banks, S. (2010). Unpublished response to Rosemary Randall's article 'Loss and Climate Change: The Cost of Parallel Narratives', published in *Ecopsychology*, September 2009.

17 Mollison, B. (1998). *Permaculture: A Designer's Manual.* Tagari Publications: Tasmania, Australia.

Chapter 2: Finding your way around local government and influencing a council

1 Wikipedia. *Local Government Act 2000.* http://en.wikipedia.org/wiki/Local_Government_Act_2000.

2 Hodgson, J. with Hopkins, R. (2010). *Transition in Action: Totnes and District 2030. An Energy Descent Action Plan.* Transition Town Totnes. An extended version of the EDAP can be viewed at www.totnesedap.org.uk.

3 Leadership Centre for Local Government (Now Local Government Leadership). *Total Place.* www.localleadership.gov.uk/totalplace.

4 Wikipedia. *Carbon Reduction Commitment.* http://en.wikipedia.org/wiki/Carbon_Reduction_Commitment.

5 Dunton, J. (2010). 'Local Carbon Frameworks pilots named'. www.lgcplus.com/lgc-news/local-carbon-framework-pilots-named/5010925.article.

6 Department of Communities and Local Government (2008). 'National Indicators for Local Authorities and Local Authority Partnerships: Handbook of Definitions'. www.communities.gov.uk/publications/localgovernment/finalnationalindicators.

7 Department of Communities and Local Government (2008). 'Key Communities, Key Resources – Engaging the capacity and capabilities of faith communities in civil resilience'. www.communities.gov.uk/publications/communities/civilresilience.

8 Transition Culture website, 27 May 2009. 'To Plan for Emergency, or Not? Heinberg and Hopkins debate'. http://transitionculture.org/2009/05/27/to-plan-for-emergency-or-not-heinberg-and-hopkins-debate.

9 Essex County Council website. 'Local banking, for Essex businesses.' www.bankingonessex.com.

10 Department of Communities and Local Government (2010). 'Listening to communities: Statutory guidance on the duty to respond to petitions'. www.communities.gov.uk/publications/communities/dutyrespondpetitionguidance.

11 Gautier, A. (2009). *Green Up! Five ways to work with your council on the environment and sustainability.* Community Development Foundation: London.

Chapter 3: The view from the inside – the experiences of four eco-councillors

1 Dean, B. (ed.) (1993). *Faith in a Seed: The Dispersion of Seeds and other late Natural History writings by Henry Thoreau.* Island Press: Washington D.C.

Chapter 4: Biodiversity and green spaces

1 *The Ecologist* (2009). 'Biodiversity "invisible" in current economic model'. www.theecologist.org/News/news_round_up/361838/biodiversity_invisible_in_current_economic_model.html.

2 Hansen, J. (2009). *Storms of my Grandchildren: The truth about the coming climate catastrophe and our last chance to save humanity.* Bloomsbury: London.

3 Connor, S. (2006). 'Earth faces "catastrophic loss of species".' *The Independent* online, 20 July 2006. www.independent.co.uk/news/science/earth-faces-catastrophic-loss-of-species-408605.html.

IUCN (2007). The IUCN Red List of Threatened Species. www.iucnredlist.org.

4 Intergovernmental Panel on Climate Change (2007). Synthesis Report of the IPCC's Fourth Assessment Report (AR4). www.ipcc.ch/publications_and_data/ar4/syr/en/spm.html.

5 IUCN (2007). The IUCN Red List of Threatened Species. www.iucnredlist.org.

6 Gedge, D. (2008, February). *Living Roofs and Walls.* www.london.gov.uk/archive/mayor/strategies/sds/docs/living-roofs.pdf.

7 Soil Association website. 'Soil carbon'. www.soilassociation.org/Whyorganic/Climatefriendlyfoodandfarming/Soilcarbon/tabid/574/Default.aspx.

8 London Wildlife Trust website. 'The Stag Beetle Project'. www.wildlondon.org.uk/Projects/TheStagBeetleProject/tabid/241/language/en-US/Default.aspx.

RSPB website. 'Big Garden Birdwatch'. www.rspb.org.uk/birdwatch.

9 Gill, S. et al. (2006). 'Adapting cities for climate change: the role of the green infrastructure'. *Built Environment*, Vol. 33, No 1. www.fs.fed.us/ccrc/topics/urban-forests/docs/Gill_Adapting_Cities.pdf.

10 Department of Communitites and Local Government (2008). 'Trees in Towns II'. www.communities.gov.uk/documents/planningandbuilding/pdf/treesintownsii.pdf.

11 Jha, A. (2010). 'Bee decline could be down to chemical cocktail interfering with brains'. *The Guardian* online, 22 June 2010. www.guardian.co.uk/environment/2010/jun/22/chemicals-bees-decline-major-study?&CMP=EMCENVEML1197.

12 Personal communication. Interviews with Farokh Khorooshi, North London Beekeepers Association, June 2010.

Chapter 5: Energy efficiency

1 Friends of the Earth (2010). *Cutting Carbon Locally and How to Pay for It.* Friends of the Earth: London.

2 Sorrell, S. (2007). *The Rebound Effect: An assessment of the evidence for economy-wide energy savings from improved energy efficiency.* UK Energy Research Centre: London.

3 Energy Saving Trust website. 'Pay As You Save Pilots'. www.energysavingtrust.org.uk/Home-improvements-and-products/Pay-As-You-Save-Pilots.

4 Energy Saving Trust (2009). 'The Energy Saving Trust's Recommendations for Planning Policies'. www.energysavingtrust.org.uk/Publication-Download/?p=1&pid=1196.

5 Building Research Establishment (2009). *Passivhaus primer.* www.passivhaus.org.uk/filelibrary/BRE-PassivHaus-Primer.pdf.

6 Hodgson, J. with Hopkins, R. (2010). *Transition in Action: Totnes and District 2030. An Energy Descent Action Plan.* Transition Town Totnes. An extended version of the EDAP can be viewed at http://totnesedap.org.uk.

7 Transition Forest Row (2009). *Forest Row in Transition.* http://there.is/TransitionForestRow-EDAP/ForestRow_In_Transition-EDAP.pdf.

Chapter 6: Energy generation

1 MacKay, D. (2008). *Sustainable Energy – Without the Hot Air.* UIT: Cambridge. Available as a free download from www.withouthotair.com.

2 Wikipedia. *District Heating.* http://en.wikipedia.org/wiki/District_heating.

3 Ibid.

4 MacKay, D. (2008). *Sustainable Energy – Without the Hot Air.* UIT: Cambridge. Available as a free download from www.withouthotair.com.

5 Wikipedia, *Solar Power.* http://en.wikipedia.org/wiki/Solar_power#Germany.

6 *The German Times*, August 2008. 'Let the sun shine on'. www.german-times.com/index.php?option=com_content&task=view&id=7692&Itemid=103.

7 Braintree Council. *Solar Water Heating*. www.braintree.gov.uk/NR/rdonlyres/F28EA33B-C282-4734-B6F5-042C5845DDFC/0/SolarWaterHeating.pdf.

8 Kirklees Council Environmental Unit (2009). 'Case Study: The RE-Charge Scheme'. www.kirklees.gov.uk/community/environment/green/pdf/RE-ChargeCaseStudyJan2010.pdf.

9 Forestry Commission England. 'Barnsley Biomass: Working towards carbon emissions reduction in Yorkshire'. www.biomassenergycentre.org.uk/pls/portal/docs/page/practical/using%20biomass%20fuels/district%20heating/barnsley%20flats.pdf.

10 Renewable UK website. 'Onshore Wind'. www.bwea.com/onshore/index.html.

11 Dafydd, I. (2009). 'Danish wind of change on energy'. BBC News Online website, 21 December 2009. http://news.bbc.co.uk/1/hi/wales/8423450.stm.

12 Hansen, J. (2009). *Storms of my grandchildren: The truth about the coming climate catastrophe and our last chance to save humanity*. Bloomsbury: London.

13 Kompogas website. www.axpo-kompogas.ch/index.php?path=home&lang=en.

14 UBM *Building* (2009). 'German city uses waste to generate green energy'. www.building.co.uk/german-city-uses-waste-to-generate-green-energy/3141605.article.

15 Kabasci, S. (2009). 'Boosting biogas with heat bonus: how combined heat and power optimizes biogas utilization'. *Renewable Energy World*, Vol. 12, Issue 5 (Sept/Oct 2009). www.renewableenergyworld.com/rea/news/article/2009/10/boosting-biogas-with-heat-bonus-how-combined-heat-and-power-optimizes-biogas-utilization.

16 Wikipedia. *Methane*. http://en.wikipedia.org/wiki/Methane.

17 Wikipedia. *Biofuel in Sweden*. http://en.wikipedia.org/wiki/Biofuel_in_Sweden.

18 Biogasmax website. 'Lille Metropolitan Area: Contribution to the Biogasmax project'. www.biogasmax.eu/biogas-techniques-biofuel-production/biogasbiofuel.html.

19 Department of Energy and Climate Change (2009). 'Biomethane into the gas network: a guide for producers'. www.decc.gov.uk/assets/decc/what%20we%20do/uk%20energy%20supply/energy%20markets/gas_markets/nonconventional/1_20091229125543_e_@@_biomethaneguidance.pdf.

20 Kaye, L. (2010). 'A Tour of Amsterdam's Waste-to-Energy Plant'. www.triplepundit.com/2010/06/a-tour-of-amsterdam%E2%80%99s-waste-to-energy-plant/.

AEB website. www.aebamsterdam.com/en/About-us/Facts-and-figures.aspx.

Chapter 7: Food

1 Lang, T. (2008). 'Food Security: Are we sleepwalking into a crisis?' City Leaders Lecture, City University London, 4 March 2008. www.city.ac.uk/aboutcity/dps/City%20Leaders%20TL%20lecture%2004%2003%2008.ppt.

2 Pinkerton, T. & Hopkins, R. (2009). *Local Food: How to make it happen in your community.* Transition Books: Dartington.

3 United Nations Food and Agriculture Organisation (2006). *Livestock's Long Shadow.* UNFAO: Rome.

4 Burnett, J. (1968) *Plenty and Want: A social history of diet in England from 1815 to the present day.* Penguin Books: Middlesex.

5 Clover, C. (2004). *The End of The Line: How overfishing is changing the world and what we eat.* Ebury Press: London.

Humphrys, J. (2001). *The Great Food Gamble.* Hodder & Stoughton: London.

Nestle, M. (2006). *What to Eat.* North Point Press: New York.

6 Hopkins, R., Thurstain-Goodwin, M. & Fairlie, S. (2010). 'Can Totnes and District Feed Itself?' http://transitionculture.org/wp-content/uploads/Can-Totnes-Feed-Itselfarticle-revised-Sept-09.pdf.

7 Designs of the time 2007 website. 'Food: the ultimate design challenge?' www.dott07.com/go/food/urban-farming.

8 Transition Town Totnes (2007). 'Totnes the nut capital of Britain' – press release. http://totnes.transitionnetwork.org/Food/TotnesTheNutCapitalOfBritain.

9 Hansen, K. & Joshi, H. (eds) (2008). Millennium Cohort

Study Third Survey. www.cls.ioe.ac.uk/text.asp?section=0001000200010011.

10 NHS (2008). 'Health Survey for England – Physical Activity and Fitness'. www.ic.nhs.uk/webfiles/publications/HSE/HSE08/HSE_08_Summary_of_key_findings.pdf.

11 National Heart Forum (2009). 'Obesity Trends for Children Aged 2-11'. www.heartforum.org.uk/images/Child_obesity_policy_long_Oct_09.pdf.

12 London Borough of Waltham Forest (2009). 'Hot Food Takeaway Supplementary Planning Document'. www.walthamforest.gov.uk/spd-hot-food-takeaway-mar10.pdf.

Chapter 8: Planning

1 Department of Communities and Local Government (2007). 'Planning Policy Statement: Planning and Climate Change. Supplement to Planning Policy Statement 1'. www.communities.gov.uk/documents/planningandbuilding/pdf/ppsclimatechange.pdf.

2 IDeA (2010). 'Dover Core Strategy evidence base'. www.idea.gov.uk/idk/core/page.do?pageId=9560694&aspect=full.

3 Hansen, J. (2009). *Storms of my Grandchildren: The truth about the coming climate catastrophe and our last chance to save humanity.* Bloomsbury: London.

4 Brighton Council (2008). *Annex to Sustainable Building Design SPD.* www.brighton-hove.gov.uk/downloads/bhcc/ldf/adopted_Sustainable_Building_Design_SPD_8_ANNEX.pdf.

Chapter 9: Procurement

1 Wikipedia. *Ponzi scheme.* http://en.wikipedia.org/wiki/Ponzi_scheme.

2 The Automatic Earth website, 17 June 2009. '40 ways to lose your future'. http://theautomaticearth.blogspot.com/2009/06/june-17-2009-40-ways-to-lose-your.html.

3 Holy Cross Time Bank website. 'Time Bank Exchange'. www.hcct.org.uk/what-we-do/timebanking/time-bank-exchange.

4 North, P. (2010) *Local Money: How to make it happen in your community.* Transition Books: Dartington.

5 Birmingham City Council (2010). Budget 2009-10. www.birmingham.gov.uk/cs/Satellite/budget?packedargs=website%3D1&rendermode=live.

6 Chantrill, C. (2010). *Local Public Spending.* www.ukpublicspending.co.uk/#ukgs302a.

7 IDeA et al. (2007). 'Sustainability and Local Government Procurement'. www.idea.gov.uk/idk/core/page.do?pageId=1707115.

8 National Sustainable Procurement Stakeholders Group. 'Sustainable Procurement Cupboard'. www.procurementcupboard.org.

9 IHS (2008). 'FAQs on EC's Green Public Procurement'. http://engineers.ihs.com/news/2008/eu-en-green-public-procurement-7-08.htm.

10 European Commission. 'Green Procurement Guide'. http://ec.europa.eu/environment/gpp/index_en.htm.

11 Forum for the Future. 'Whole Life Costing (+CO_2) User Guide'. www.forumforthefuture.org.uk/files/WLC-CO2-tool-user-guide-WEB.pdf.

Chapter 10: Recycling

1 Hosking, R. (2007). 'Carrying conviction'. *The Guardian* online, 16 May 2007. www.guardian.co.uk/society/2007/may/16/business.waste.

2 www.plasticbagfree.com (accessed 1 April 2010 but site not available when this book went to print).

3 letsrecycle.com, 27 April 2007. 'Trading standards call for tougher laws on excess packaging'. www.letsrecycle.com/do/ecco.py/view_item?listid=37&listcatid=225&listitemid=8684§ion=materials/packaging.

4 Boyle, D. (2009). *Greening Actually.* Local Government Association: London.

5 Bath and North East Somerset Council website. 'Zero Waste'. www.bathnes.gov.uk/BathNES/environmentandplanning/recyclingandwaste/zerowaste.htm.

6 Big Brother Watch website, 3 April 2010. 'Lifting the Lid: Number of councils installing chips in bins rises by 62 per cent'. www.bigbrotherwatch.org.uk/home/2010/03/

lifting-the-lid-number-of-councils-installing-chips-in-bins-rises-by-62-per-cent.html.

7 BBC News Online website, 1 June 2010. 'Windsor recycling reward scheme adopted by government'. news.bbc.co.uk/1/hi/england/berkshire/10253576.stm.

8 Mariën, L. (2009). 'Prevention and Management of Household Waste in Flanders'. www.foe.co.uk/resource/event_presentations/2_lore_marien.pdf.

9 BBC News Online website, 19 June 2010. 'Re-useable items salvaged and sold at Hampshire tips'. http://news.bbc.co.uk/1/hi/uk/10356725.stm.

10 London Borough of Camden (2008). 'Energy Audit of the Kerbside Recycling Service'. www.camden.gov.uk/ccm/content/environment/policies-reports-and-data/energy-audit-of-the-kerbside-recycling-services.en;jsessionid=300770C1E81E37828DBFC6C2D68E76BF.

11 WRAP website. 'About food waste'. www.lovefoodhatewaste.com/about_food_waste.

Chapter 11: Transport

1 Sustrans website. 'DIY Streets'. www.sustrans.org.uk/what-we-do/liveable-neighbourhoods/diy-streets.

2 National Travel Survey. www.statistics.gov.uk/ssd/surveys/national_travel_survey.asp.

3 Sustrans et al. (2008). 'Take Action on Active Travel: Why a shift from car-dominated transport policy would benefit public health'. www.adph.org.uk/downloads/policies/Take_action_on_active_travel.pdf.

4 US Federal Transport Administration website, 13 Jan 2010. 'Obama Administration proposes major public transportation policy shift to highlight livability'. www.fta.dot.gov/news/news_events_11036.html.

5 Ashton, S. & Mackay, G. (1979). 'Some characteristics of the population who suffer trauma as pedestrians when hit by cars and some resulting implications'. Proceedings of the IRCOBI Conference, Gothenburg, Sweden.

6 Transport Research Laboratory (1996). 'Review of Traffic Calming Schemes in 20mph Zones'. www.trl.co.uk/online_store/reports_publications/trl_reports/cat_traffic_engineering/report_review_of_traffic_calming_schemes_in_20_mph_zones.htm.

7 Smarter Travel Sutton website, 8 February 2010. 'Pioneering green travel programme delivers 75 per cent increase in cycling'. www.smartertravelsutton.org/home.

8 Hillman, M. (ed.) (1993). *Children, Transport and the Quality of Life*. Policy Studies Institute. www.psi.org.uk/mayerhillman/Children%20Transport%20Quality%20of%20Life.pdf.

HM Treasury website, 25 March 2010. 'Total Place'. www.hm-treasury.gov.uk/psr_total_place.htm.

9 Institute for European Environmental Policy (2007). 'Unfit for Purpose: How Car Use Fuels Climate Change and Obesity'. www.ieep.eu/publications/pdfs/2007/IEEP%20-%20Unfit%20for%20purpose_transport%20climate%20chage%20and%20obesity.pdf.

10 Wikipedia, *Road pricing*. http://en.wikipedia.org/wiki/Road_pricing.

11 Friends of the Earth (2010). *Cutting Carbon Locally and How to Pay for It.* Friends of the Earth: London.

12 Robinson, M. (2010). 'TfL attacked for transporting 'green' buses around the world'. *London Evening Standard* online, 24 March 2010. www.thisislondon.co.uk/standard/article-23818319-tfl-attacked-for-transporting-green-buses-around-the-world.do

13 Brighton Council (2009). 'Local Transport Plan Delivery Report'. www.brighton-hove.gov.uk/downloads/bhcc/LTP_DR.pdf.

14 Adelaide City Council website. 'Tindo – the world's first solar electric bus'. www.adelaidecitycouncil.com/adccwr/publications/guides_factsheets/tindo_fact_sheet.pdf.

15 Kelly, F. & Birkett, S. (2009). 'Presentations given to Camden Council's Sustainability Task Force, 22 April 2009'. www.camden.gov.uk/ccm/cms-service/stream/asset/?asset_id=1668343.

16 Sustrans website. 'Nine thousand voices for safer cycling'. www.sustrans.org.uk/resources/in-the-news/nine-thousand-voices-for-safer-cycling.

17 Cycling England (2007). 'Bike for the Future II'. www.dft.gov.uk/cyclingengland/site/wp-content/uploads/2008/08/bike-for-the-future-ii.pdf.

18 Monbiot, G. (2006). *Heat: How to stop the planet burning.* Allen Lane (Penguin Group): London.

Chapter 12: Water

1 Le Quesne, T. et al. (2007). *Allocating Scarce Water: A World Wildlife Fund primer on water allocation, water rights and water markets.* WWF: Godalming, UK.

2 Pearce, F. (2006). *When The Rivers Run Dry.* Transworld Publishers (Random House Group): London.

3 Waterwise website. 'The Facts'. www.waterwise.org.uk/reducing_water_wastage_in_the_uk/the_facts/the_facts_about_saving_water.html.

4 Met Office website. 'What does it mean for the world?' www.metoffice.gov.uk/climatechange/guide/quick/impacts.html.

5 Bates, B. et al. (2008). 'Climate Change and Water – Technical Paper of the Intergovernmental Panel on Climate Change'. www.ipcc.ch/pdf/technical-papers/ccw/frontmatter.pdf.

6 Ibid.

7 United Nations Population Fund (2007). *State of the World Population 2007.* UNFPA: New York.

8 Evans-Pritchard, A. (2008). 'Water crisis to be biggest world risk'. *The Telegraph* online, 5 June 2008. www.telegraph.co.uk/finance/newsbysector/utilities/2791116/Water-crisis-to-be-biggest-world-risk.html.

9 Renault, D. & Wallender, W. (2000). 'Nutritional water productivity and diets'. *Agricultural Water Management*, Vol. 45, No.3 (August 2000), 275-96.

10 Chapagain, A. & Orr, S. (2008). *UK Water Footprint: The impact of the UK's food and fibre consumption on global water resources.* WWF-UK: Godalming, Surrey.

11 Hoekstra, A. & Hung, P. (2002). *Virtual Water Trade: A quantification of virtual water flows between nations in relation to international crop trade.* Value of Water Research Report Series No. 11, Institute for Water Education, UNESCO. Delft, The Netherlands: UNESCO-IHE.

12 Chapagain, A. & Hoekstra, A. (2007). 'Water footprints of nations: Water use by people as a function of their consumption pattern'. *ISESCO Science and Technology Vision,* Vol. 4, No.5 (May 2008), 38-42.

13 National Grid website. *'National Grid powers up for World Cup 2010'.* www.nationalgrid.com/uk/Media+Centre/WorldCup2010.

14 Personal communication. Lutz Johnen, Aquality. Interviewed May 2008.

15 Gillan, A. & Morris, S. 'Thousands without fresh water as floods bring chaos'. *The Guardian* online, 23 July 2007. www.guardian.co.uk/uk/2007/jul/23/weather.immigrationpolicy.

16 Mayor of London (2009). 'The Mayor's Draft Water Strategy'. http://static.london.gov.uk/mayor/environment/water/docs/draft-water-strategy.pdf.

17 Department of Environment, Food and Rural Affairs (2008). 'Future Water'. www.defra.gov.uk/environment/quality/water/strategy/pdf/future-water.pdf.

18 Hampshire County Council (2000). 'Water in Hampshire: A comprehensive review'. www.ecoreports.co.uk/media/otherreports/report23.pdf.

Chapter 13: Well-being

1 Bacon, N. et al. (2010). 'The State of Happiness: Can public policy shape people's wellbeing and resilience?' www.youngfoundation.org/files/images/wellbeing_happiness_Final_webversion.pdf.

2 Jackson, T. (2009). *Prosperity without Growth.* Earthscan: London.

3 Layard, R. (2005). *Happiness.* Penguin: London.

4 new economics foundation website. 'National Accounts of Well-being'. www.nationalaccountsofwellbeing.org/learn/related/five-ways-start.html.

5 Bristol City Council website. 'Indicators of the quality of life in Bristol'. www.bristol.gov.uk/ccm/content/Council-

Democracy/Statistics-Census-Information/indicators-of-the-quality-of-life-in-bristol.en.

Manchester City Council. 'Community Guardian'. www.challengemanchester.co.uk/guardians.

new economics foundation website, 27 April 2004. 'The power and potential of well-being indicators: measuring young people's well-being in Nottingham'. www.neweconomics.org/publications/power-and-potential-well-being-indicators.

Thanet District Council website. 'Safer Stronger Community Fund'. www.thanet.gov.uk/business/regeneration/safer_stronger_communities.aspx.

6 Bacon, N. et al. (2010). 'The State of Happiness: Can public policy shape people's wellbeing and resilience?' www.youngfoundation.org/files/images/wellbeing_happiness_Final_webversion.pdf.

Chapter 14: Getting elected

1 Transition Network (2010). 'How Resilient Are Our Political Candidates? Embedding Transition in the Political Process. A Hustings Primer from the Transition Network 2010'. http://transitionculture.org/wp-content/uploads/Final-designed-hustings-version-2.0.pdf.

2 Ibid.

3 Seyfang, G. (2010). 'Transition Norwich: A fine city in Transition. Report of the 2009 Membership Survey'. http://transitionculture.org/wp-content/uploads/Transition-Norwich-2009-Survey-Report.pdf.

RESOURCES

This section gives details of the organisations, websites and publications referred to in this book, as well as a number of others relating to the issues covered.

Contents

Peak oil and energy transition

Publications

APPGOPO (2009). *Tradable Energy Quotas (TEQs): A Policy Framework For Peak Oil and Climate Change.* All Party Parliamentary Group on Peak Oil and the Lean Economy Connection.

City of Portland Peak Oil Task Force (2007). 'Descending the Oil Peak: Navigating the Transition from Oil and Natural Gas'. www.portlandonline.com/bps/index.cfm?a=145732&c=42894.

Heinberg, R. (2007). *The Party's Over: Oil, war and the fate of industrial societies.* Clairview Books.

Heinberg, R. (2007). *Peak Everything: Waking up to the century of decline in Earth's resources.* Clairview Books.

Heinberg, R. (2007). *Powerdown: Options and actions for a post-carbon society.* Clairview Books.

ITPOES (2010). 'The Oil Crunch: A wake-up call for the UK economy. Second report of the UK Industry Task Force on Peak Oil and Energy Security'. Industry Taskforce on Peak Oil and Energy Security. http://peakoiltaskforce.net/wp-content/uploads/2010/02/final-report-uk-itpoes_report_the-oil-crunch_feb20101.pdf.

Sorrell, S. et al. (2009) 'Global Oil Depletion: An assessment of the evidence for a near-term peak in global oil production'. UK Energy Research Centre.

Spratt, S. et al. (2009). *The Great Transition: A tale of how it turned out right.* new economics foundation.

Strahan, D. (2008). *The Last Oil Shock: A survival guide to the imminent extinction of Petroleum Man.* John Murray.

Walker, B. & Salt, D. (2006) *Resilience Thinking: Sustaining ecosystems and people in a changing world.* Island Press.

Organisations and websites

All Party Parliamentary Group on Peak Oil and Gas (APPGOPO)
020 7219 6314
www.appgopo.org.uk
APPGOPO investigates and discusses the debate about global peak oil production and looks at the impacts, mitigations and solutions.

Association for the Study of Peak Oil (ASPO)
www.peakoil.net
ASPO is a network of scientists and others working to determine the date and impacts of the peak and decline in world oil and gas production.

Oil Depletion Analysis Centre (ODAC)
020 8144 8359
www.odac-info.org
ODAC works to raise international awareness and promote a better understanding of the world's oil depletion problem.

Post Carbon Institute (PCI)
www.postcarbon.org
The PCI provides the resources needed to understand and respond to economic, energy and environmental crises and so enables the move towards resilient communities and relocalised economies.

UK Industry Taskforce on Peak Oil and Energy Security (ITPOES)
020 7401 8001
http://peakoiltaskforce.net
A group of British companies concerned that inadequate attention is being paid to the threats to energy security.

http://postcarboncities.net
Post Carbon Cities. This website provides resources for local governments on peak oil and global warming.

Climate change and carbon-cutting

Publications

Department of Trade and Industry (2003). *Our Energy Future: Creating a low-carbon economy.* The Stationery Office.

Hamilton, C. (2010). *Requiem for a Species: Why we resist the truth about climate change.* Earthscan.

Hansen, J. (2009). *Storms of my Grandchildren: The truth about the coming climate catastrophe and our last chance to save humanity.* Bloomsbury.

Lynas, M. (2008). *Six Degrees: Our future on a hotter planet.* Harper Perennial.

Middlemiss, L. & Parrish, B. D. (2010). 'Building capacity for low carbon communities: the role of grassroots initiatives'. *Energy Policy*, in press. doi:10.1016/j.enpol.2009.07.003.

Organisations and websites

10:10 Campaign
020 7388 6688
www.1010global.org/uk
The 10:10 Campaign aims to unite every sector of society behind a simple idea: to cut our emissions by 10% in 2010.

BTCV Carbon Army
www2.btcv.org.uk/display/carbonarmy
The Carbon Army is BTCV's practical hands-on response to climate change. It tackles the impacts of climate change through planting trees, managing water resources to reduce risks of floods, and create habitats for endangered native species.

Carbon Conversations
01223 971353
www.cambridgecarbonfootprint.org/action/carbon-conversations/national-carbon-conversations
A practical and inspiring six-session course on low-carbon living based on the psychology of change. Initially available only in Cambridge, Carbon Conversations is now available for use elsewhere after completing the two-day facilitator training.

Carbon Disclosure Project (CDP)
020 7970 5660
www.cdproject.net/en-US/Pages/HomePage.aspx
Thousands of organisations disclose their greenhouse gas emissions and climate change strategies through CDP, so it holds the largest database of primary corporate climate change information in the world. This information is put at the heart of financial and policy decision-making.

Carbon Emissions Reduction Target scheme (CERT)
www.decc.gov.uk/en/content/cms/what_we_do/consumers/saving_energy/cert/cert.aspx
CERT requires domestic energy suppliers with over 50,000 customers to reduce the carbon emissions of householders, focusing particularly on vulnerable and low-income households by increasing the energy efficiency of their homes.

Carbon Rationing Action Groups (CRAGS)
www.carbonrationing.org.uk
CRAGs implement carbon emissions reductions at the community level by forming local groups to encourage one another to reduce carbon footprints.

Carbon Trust (and its local government programme)
0800 085 2005
www.carbontrust.co.uk/Pages/Default.aspx
Helps businesses, the public sector and local authorities to cut carbon emissions, save energy and commercialise low-carbon technologies. Provides local authorities with advice, resources, case studies, a network for exchange of experiences and information, and opportunities for collaboration.

Committee on Climate Change (CCC)
020 7592 1553
www.theccc.org.uk
The CCC advises the UK government on setting carbon budgets, and reports to Parliament on progress in greenhouse gas emissions reductions.

CRC Energy Efficiency Scheme
(formerly the Carbon Reduction Commitment)
www.carbonreductioncommitment.info
A mandatory carbon-trading scheme aimed at energy saving and carbon emissions reduction.

Intergovernmental Panel on Climate Change (IPCC)
www.ipcc.ch
Assesses climate change and provides a clear scientific view on its current state and the potential environmental and socio-economic consequences.

Tyndale Centre for Climate Change
01603 593900
www.tyndall.ac.uk
Brings together scientists, economists, engineers and social scientists who are working to develop sustainable responses to climate change. Also works with business leaders, policy advisors, the media and the general public.

UK Climate Impacts Programme (UKCIP)
01865 285717
www.ukcip.org.uk
Helps organisations adapt their work and activities to the impacts of climate change.

www.energysavingtrust.org.uk/nottingham
The Nottingham Declaration website. Provides climate change information and advice for English local authorities. By signing the Nottingham Declaration, councils and their partners pledge to systematically address the causes of climate change and to prepare their communities for its impacts. The action pack provides guidance for producing mitigation and adaptation Action Plans.

www.realclimate.org
The best place for up-to-the-minute news about climate change, as posted by climate scientists.

www.teqs.net
Tradable Energy Quotas is a system that would attempt to ration energy use and cut nations' carbon emissions along with their use of oil, gas and coal and, its proponents suggest, ensure fair access to energy for all. This website explains how.

The Transition movement

Publications – general

Bird, C. (2010). *Local Sustainable Homes: how to make them happen in your community*. Transition Books.

Chamberlin, S. (2009).*The Transition Timeline: For a local, resilient future*. Green Books.

Hopkins, R. (2008). *The Transition Handbook: From oil dependency to local resilience*. Green Books.

North, P. (2010). *Local Money: how to make it happen in your community*. Transition Books.

Pinkerton, T. & Hopkins, R. (2009). *Local Food: How to make it happen in your community*. Transition Books.

Seyfang, G. (2009) *Green Shoots of Sustainability: The 2009 Transition movement survey*. University of East Anglia.

Transition Network (2010). 'How Resilient Are Our Political Candidates? Embedding Transition in the Political Process. A Hustings Primer from the Transition Network 2010'. Totnes, Transition Network. www.transitionculture.org/wp-content/uploads/Final-designed-hustings-version-2.0.pdf.

Energy Descent Action Plans

Bristol City Council. 'Peak Oil Report: Building a positive future for Bristol after peak oil'. www.bristol.gov.uk/ccm/content/Environment-Planning/sustainability/file-storage-items/peak-oil-report.en.

Hodgson, J., with Hopkins, R. (2010). *Transition in Action: Totnes and District 2030: An Energy Descent Action Plan*. Transition Town Totnes.
The UK's first thorough EDAP. An extended version of the EDAP can be viewed at http://totnesedap.org.uk.

Sustainable Frome (2009). *Sustainable Frome: A town in Transition. Energy Descent Action Plan* [work in progress].

Transition Forest Row (2009). *Forest Row In Transition*. http://there.is/TransitionForestRow-EDAP/ForestRow_In_Transition-EDAP.pdf.

Organisations and websites

The Transition Network
05601 531882
www.transitionnetwork.org

www.transitionnetwork.org/initiatives
A directory of all the official and many of the mulling Transition Initiatives across the world.

www.transitionbooks.net
Transition Books, an imprint of Green Books.

www.transitionculture.org
Rob Hopkins' blog – 'an evolving exploration of the head, heart and hands of energy descent'.

Local government and communities

Publications

Centre for Community Enterprise (2000). *The Community Resilience Manual: A resource for rural recovery and renewal.* Centre for Community Enterprise.

DEFRA (2010). 'Adapting to Climate Change: A guide for local councils'. Department for the Environment, Food and Rural Affairs.

Godfrey, C. & Birch, P. (2009). 'Towards a Resilient Taunton Deane: 375 Voices, One Story'. Taunton Deane Borough Council and Transition Town Taunton.

Hurring, D. (2009). 'Transition Audit of Somerset County Council'. Operations Director, Natural Communities Community Interest Company.

Local Government Association (LGA) (2008). 'Be Aware, Be Prepared, Take Action: How to integrate climate change adaptation strategies into local government'. Local Government Association and Environment Agency. www.lga.gov.uk/lga/aio/566302.
This report includes useful tools and resources that local authorities can use when developing climate change adaptation strategies.

Local Government Association (LGA) (2008). 'Volatile Times: transport, climate change and the price of oil'. www.lga.gov.uk/lga/aio/1335142.
This report analyses the evidence and theory surrounding the debate on oil price variability. It recommends practical steps for adapting councils, businesses and householders to rising oil prices; gives signposts to further support; and provides case study materials.

McDonald, N. (2010). 'The role of Transition Initiatives in local authorities' responsiveness to peak oil: A case study of Somerset County Council'. www.transitionculture.org/wp-content/uploads/09_09_15-DISSERTATION-FINAL-VERSION.pdf.
A well-researched, detailed account of Somerset County Council's resolution.

Nottingham City Council (2008). Peak Oil Resolution: passed 8 December 2008. http://open.nottinghamcity.gov.uk/comm/download3.asp?dltype=inline&filename=36153/081208.doc.

ODAC (2007) 'Preparing for Peak Oil: Local Authorities and the Energy Crisis'. Oil Depletion Analysis Centre/Post Carbon Institute. www.odac-info.org/sites/default/files/Preparing_for_Peak_Oil_0.pdf.
This report is aimed specifically at local government in the UK. It summarises what local authorities are doing to tackle peak oil and which policies are the most promising so that all can benefit from these best practices.

Pitts, R. (2009) 'Peak Oil and Energy Uncertainty: Information for local authorities'. Welsh Local Government Association's Sustainable Development Framework. www.wlga.gov.uk/english/archive-of-reports9/peak-oil-and-energy-uncertainty.

Organisations and websites

Community Development Foundation (CDF)
020 7833 1772
www.cdf.org.uk
A public body and a charity, the CDF provides a bridge between government, communities and the voluntary sector.

Department of Communities and Local Government
www.communities.gov.uk
Promotes decentralisation and democratic engagement to end top-down government by giving new powers to councils, communities, neighbourhoods and individuals.

IDeA
See Local Government Improvement and Development (below)

Local Government Improvement and Development
(formerly the Improvement and Development Agency – IDeA)
020 7296 6880
www.idea.gov.uk
Supports improvement and innovation in local government through networks, online communities of practice and web resources.

Local Government Association (LGA)
020 7664 3000
www.lga.gov.uk
The voice of local government in the national arena. The LGA lobbies and campaigns for changes in policy, legislation and funding on behalf of its member councils and the people it serves.

National Association of Local Councils (NALC)
020 7637 1865
www.nalc.gov.uk
NALC represents the interests of town and parish councils in England, and works with and for its member councils by lobbying national government to advance and protect those interests.

www.cabinetoffice.gov.uk/ukresilience.aspx
Government resilience website. Provides advice on emergency response and recovery, and resources for civil-protection practitioners to improve emergency preparedness.

Environment

Organisations and websites

Campaign to Protect Rural England (CPRE)
020 7981 2800
www.cpre.org.uk
Campaigns for a sustainable future for the English country-side, highlights threats and promotes positive solutions.

Environment Agency
0870 850 6506
www.environment-agency.gov.uk
Protects and improves the environment and promotes sustainable development, and delivers the environmental priorities of central government and the Welsh Assembly.

European Environmental Policy Institute
www.ieep.eu
An independent not-for-profit institute dedicated to advancing an environmentally sustainable Europe through policy analysis, development and dissemination.

Greenspace Information for Greater London
www.gigl.org.uk
GiGL is London's open space and biodiversity records centre, where detailed information on London's wildlife, parks, nature reserves, gardens and other open spaces is collated, managed and made available.

London Wildlife Trust Green Schools
020 7261 0447
www.wildlondon.org.uk/Education/Servicesforschools/tabid/258/language/en-GB/Default.aspx
Provides teaching packs for effective use of outdoor spaces (e.g. wildlife gardening), education sessions in schools, advice for using school grounds, and events organisation.

Marine Conservation Society (MCS)
01989 566017
www.mcsuk.org
Champions marine wildlife protection, sustainable fisheries

to support future generations, and clean seas and beaches for people and wildlife.

Species to avoid: www.fishonline.org/advice/avoid.
List of fish that MCS believes are most vulnerable to over-fishing and/or are being caught using methods that damage the environment or non-target species.

Species to eat: www.fishonline.org/advice/eat.
List of fish that MCS believes are fished within sustainable levels using methods that don't cause unacceptable damage to the environment or non-target species.

Pesticide Action Network (PAN)
020 7065 0905
www.pan-uk.org
PAN publishes independent information on pesticide use and impacts, and researches causes and effects of pesticide problems.

Wildlife Trusts
01636 677711
www.wildlifetrusts.org
The Wildlife Trusts manage thousands of nature reserves and marine conservation projects, advise landowners on wildlife-friendly land management and also work with schools.

http://transitionculture.org/2010/07/21/jeremy-jackson-on-how-we-wrecked-the-oceans
A TED talk by Jeremy Jackson on how we ruined the oceans.

Energy efficiency and energy generation

Publications

Dunning, J. & Turner, A. (2005). 'Community-owned windfarms – aspirations, suspicions and reality'. *PowerUK*, 131, 42-45.

Kemp, M. & Wexler, J. (eds) (2010). 'Zero Carbon Britain 2030: A new energy strategy. The second report of the Zero Carbon Britain project'. Centre for Alternative Technology, Machynlleth.

MacKay, D. (2008). *Sustainable Energy – Without the Hot Air.* UIT Press. Available as a free download from www.withouthotair.com.

Walker, G., Devine-Wright, P. & Evans, B. (2007). 'Community Energy initiatives: Embedding sustainable technologies at a local level'. End of Award Report. Economic and Social Research Council. http://geography.lancs.ac.uk/cei/index.htm.

Organisations

Energy Saving Trust
0800 512012
www.energysavingtrust.org.uk
Promotes actions that reduce carbon emissions and provides free advice and information for people who want to save energy, conserve water and reduce waste.

Energy4All
www.energysteps.coop
01229 821028
Energy4All helps communities assess whether their wind energy project could be successful, and then offers support to groups that have the drive and commitment to deliver a community-owned energy project.

Government Feed-in Tariff
0800 512012
www.energysavingtrust.org.uk/Generate-your-own-energy/Sell-your-own-energy/Feed-in-Tariff-Clean-Energy-Cashback-scheme
Under this scheme, energy suppliers make regular payments to householders and communities who generate their own electricity from renewable or low-carbon sources such as solar panels or wind turbines.

Home Energy Efficiency for Tomorrow
(formerly the Home Energy Efficiency Programme)
020 7934 9776
www.londoncouncils.gov.uk/capitalambition/projects/homesenergyefficiencyprogramme.htm
A pan-London retrofitting scheme for reducing residential carbon emissions. It will begin with trials and demonstration projects for practical energy efficiency and the refined model will eventually be rolled out across London.

Kirklees Warm Zone
www.warmzones.co.uk/kirklees.html
The Warm Zones project (see overleaf) for Kirklees.

National Energy Action (NEA)
0191 261 5677
www.nea.org.uk
The NEA aims to eradicate fuel poverty by developing and promoting energy efficiency services to tackle heating and insulation problems of low-income households, and by campaigning for more investment in energy efficiency to help people who are poor and vulnerable.

Sustainable Energy Academy
www.sustainable-energyacademy.org.uk
Promotes education and action to reduce the carbon footprints of homes. The Sustainable Energy Academy is building a network of 'Old Home SuperHome' as exemplars of old dwellings retrofitted for energy efficiency and accessible to the public.

UK Energy Research Centre (UKERC)
020 7594 1574
www.ukerc.ac.uk
UKERC researches sustainable future energy systems and informs UK policy development and research strategy.

Warm Zones
www.warmzones.co.uk
Warm Zones develops new approaches to fuel poverty by delivering practical and cost-effective means of improved energy efficiency as well as contributing to economic and social regeneration.

Building

Publications

Building Research Establishment. *PassivHaus Primer.* www.passivhaus.org.uk/filelibrary/BRE-PassivHaus-Primer.pdf.

Department of Communities and Local Government (2008). 'The Code for Sustainable Homes: Setting the standard in sustainability for new homes'. www.communities.gov.uk/publications/planningandbuilding/codesustainability standards
The Code for Sustainable Homes is an environmental assessment for new-build residential developments that awards points on a number of environment principles. This document sets out the assessment process and the performance standards required for the Code.

Hall, K. (2009). 'Passive house systems arrive in the UK'. *Green Building* magazine, 19(1) (Summer 2008).

Pickerill, J. & Maxey, L. (eds) (2008). *Low Impact Development: The future in our hands.* University of Leicester Department of Geography. Available at www.lowimpactdevelopment.wordpress.com.

Williamson, J. (2009.) 'First certified domestic PassivHaus in the UK'. *Green Building* magazine, 19(2) (Autumn 2009).

Organisations

Amazonails
www.amazonails.org.uk
0845 458 2173 or 01706 814696
The leading UK company in straw-bale building. Offers design, consultancy, training and support in straw-bale and other sustainable building techniques.

PassivHaus UK
0845 873 5552
www.passivhaus.org.uk
Passivhaus is a specific construction standard for buildings, and applies to homes as well as commercial, industrial and public buildings. Buildings typically achieve an energy saving of 90% compared with typical housing.

Sustainable Building Association (AECB)
0845 456 9773
www.aecb.net
A network of individuals and companies who develop, share and promote best practice in environmentally sustainable building.

Food

Publications

Heinberg, R. & Bomford, M. (2009). *The Food and Farming Transition: Towards a post-carbon food system.* Post Carbon Institute.

Hopkins, R., Thurstain-Goodwin, M. & Fairlie, S. (2009). *Can Totnes and District Feed Itself? Exploring the practicalities of food relocalisation. Working Paper Version 1.0.* Transition Town Totnes / Transition Network.

Morgan, K. & Sonnino, R. (2008). *The School Food Revolution: Public food and the challenge of sustainable development.* London, Earthscan.

Transition Norwich (2009). *Outline of a Food Chapter for the Energy Descent Plan for Norwich.* Transition Network / East Anglia Food Links.

Transition Stroud (2008). *Food Availability in Stroud District: Considered in the context of climate change and peak oil.* For the Local Strategic Partnership Think Tank on Global Change.

Viljoen, A. (ed.) (2005). *Continuous Productive Urban Landscapes: Designing Urban Agriculture for Sustainable Cities.* Architectural Press.

Organisations and websites – general

Agroforestry Research Trust
01803 840776
www.agroforestry.co.uk
Researches temperate agroforestry and all aspects of cropping and uses. It also produces publications, and sells plants and seeds.

Federation of City Farms and Community Gardens
0117 923 1800
www.farmgarden.org.uk
Supports, represents and promotes community-managed farms, gardens, allotments and other green spaces. Works with people from all ages, backgrounds and abilities to build better communities and environments, often in deprived areas.

Garden Organic
024 7630 3517
www.gardenorganic.org.uk
Garden Organic is dedicated to researching and promoting organic gardening, farming and food.

Permaculture Association
0845 458 1805
www.permaculture.org.uk
The Permaculture Association provides advice, information, support and training in the theory and practice of permaculture.

Soil Association
www.soilassociation.org
0117 314 5000
Promotes planet-friendly food and farming through education, campaigns and community programmes.

The Soil Association's Community Supported Agriculture programme:
www.soilassociation.org/Takeaction/Getinvolvedlocally/Communitysupportedagriculture/tabid/201/Default.aspx
Provides a network, case studies and information about training events to help farmers and communities start, develop and join a CSA.

www.meatfreemondays.co.uk
Meat Free Mondays website. Encourages people to have a meat-free day each week to improve their health, reduce the environmental impacts of food production, and as a way of combating global warming.

Land-share and garden-share schemes

Grow Your Neighbour's Own
http://grow.transitionbrightonandhove.org.uk
A local landshare project for the Brighton and Hove area, which pairs growers with garden- and land-owners.

LandShare
www.landshare.org
A social enterprise project that aims to be a catalyst for change in the way land and resources are managed. It connects policymakers, farmers, foresters and academics, and implements practical projects.

Landshare
www.landshare.net
Emerging from the TV series *River Cottage*, Landshare brings together people who want to grow food but don't have any land with people who have spare land and are willing to share it.

Transition Cambridge Garden Share Scheme
www.transitioncambridge.org/thewiki/ttwiki/pmwiki.php?n=TTFood.GardenShare
Another local landshare scheme, this one matches owners and growers by location and expertise (i.e. beginners with experts).

Transition Town Totnes Garden Share Project
01803 867358
http://totnes.transitionnetwork.org/gardenshare/home
A local project linking people who have unused patches of garden with local people who want to grow food but have no access to land.

Procurement

Publications

European Commission. *Buying Green! A handbook on environmental public procurement.* http://ec.europa.eu/environment/gpp/buying_handbook_en.

Forum for the Future. 'Whole Life Costing (+CO_2) User Guide'. www.forumforthefuture.org.uk/files/WLC-CO2-tool-user-guide-WEB.pdf
This guide is for procurement professionals who evaluate different product options given by suppliers, and will help in understanding which product will result in fewer carbon emissions over its entire lifetime.

Hulme, J. & Radford, N. (2010). *Sustainable Supply Chains That Support Local Economic Development.* Prince's Foundation for the Built Environment.

Organisations and websites

Corporate Assessment of Environmental, Social and Economic Responsibility (CAESER)
www.caeser.org
0845 299 2994
CAESER is a supplier assurance programme primarily adopted by public sector procurement specialists. It provides a structured approach to engaging suppliers on key areas of the government's sustainability agenda.

European Commission

Green Public Procurement Helpdesk:
00 800 6789 1011, http://ec.europa.eu/environment/gpp/index_en.htm
Helpdesk that disseminates information on GPP and answers stakeholders' enquiries. Phone line is free from anywhere in the 27 EC Member States.

Green Public Procurement Training Kit:
http://ec.europa.eu/environment/gpp/toolkit_en.htm
The GPP's Training Kit has three modules: an Action Plan for GPP, a Legal Module and a Practical Module.

www.procurementcupboard.org
The Sustainable Procurement Cupboard provides professionals with case studies, tools and information to help them implement sustainable procurement practices in their organisations.

Resource use and recycling

Organisations and websites

Campaign for Real Recycling
www.realrecycling.org.uk
This campaign is pushing government and local authorities to improve the quality of collected recyclables by implementing highly separated recycling systems.

ECT Recycling
020 8813 3880
www.ectrecycling.co.uk
One of the UK's leading community recycling schemes.

Waste and Resources Action Plan (WRAP)
01295 819900
www.wrap.org.uk
Helps people and businesses to reduce waste, develop sustainable products and use resources in an efficient way.

www.lovefoodhatewaste.com/about_food_waste:
WRAP's 'love food hate waste' campaign, which raises awareness of the need to reduce food waste and how this can benefit consumers and the environment.

www.ilovefreegle.org
Freegle. A Yahoo! Groups email list, which, like Freecycle (see opposite, top), enables people to give stuff away when they need to get rid of it, or find something they need that may otherwise be thrown away.

www.instructables.com
An online community where members exchange instructions on how to produce home-made alternatives to consumer goods.

www.uk.freecycle.org
Freecycle. A worldwide network matching people who have things they no longer want with people in their own towns who can make use of them. The goal is to keep usable items out of landfills.

Transport

Publications

Engwicht, D. (2005). *Mental Speed Bumps: The smarter way to tame traffic*. Envirobook. www.mentalspeedbumps.com.

Gilbert, R. & Perl, A. (2010). *Transport Revolutions: Moving people and freight without oil* (2nd edn). New Society Publishers.

Sloman, L. (2006) *Car Sick: Solutions for our car-addicted culture*. Green Books.

Organisations and websites

Campaign for Better Transport
020 7566 6480
www.bettertransport.org.uk
This campaign pushes for innovative practical policies and transport solutions that improve people's lives and reduce environmental damage.

Commonwheels
0845 602 8030
www.commonwheels.org.uk
Commonwheels is a company that offers local car clubs on a 'pay as you drive' basis.

Do It Yourself Streets
www.sustrans.org.uk/what-we-do/liveable-neighbour-hoods/diy-streets
Sustrans' DIY Streets brings communities together to redesign their streets in a way that puts people at their heart and makes them safer and more attractive places to live.

DIY Streets information sheets:
www.sustrans.org.uk/assets/files/liveable20neighbour-hoods/DIY20Streets20info20sheet_FINAL.pdf

DIY Streets Guide:
http://www.sustrans.org.uk/assets/files/liveable20neighbourhoods/A20simple20guide.pdf

Sustrans
0845 113 00 65
www.sustrans.org.uk
The UK's leading sustainable transport charity. It works on practical innovative ways of dealing with the transport challenges we face.

www.carplus.org.uk/car-clubs/find-a-car-club-car
This site is the place to go if you want to find car clubs operating in your area.

Water

Publications

Bathroom Manufacturers Association. 'Let's Use Water Wisely'. www.bathroom-association.org/waterhog/factsheet.asp
Useful information on devices that save water.

Organisations and websites

Consumer Council for Water
0121 345 1000 or 0845 039 2837
www.ccwater.org.uk
Represents the interests of water consumers across England and Wales. Its factsheet 'Why Tap Water is a Winner' is available at www.ccwater.org.uk/upload/pdf/why_tap_water_winner.pdf.

Waterwise
020 7344 1882
www.waterwise.org.uk
A UK NGO that encourages decreased water consumption in the UK and builds evidence for large-scale improved water efficiency.

Westcountry Rivers Trust (WRT)
01579 372140
www.wrt.org.uk
Works for the preservation, protection, development and improvement of the West Country's rivers, streams, watercourses and water impoundments, and educates people on water management.

www.bathroom-association.org/waterhog
Bathroom Manufacturers Association Waterhog Site. General water-saving advice and a search facility for manufacturers that make water-efficient products.

www.footprint-trust.co.uk/water_works.htm
The Footprint Trust's 'Water Works' initiative, which educates people on the need to use water resources more efficiently.

Sustainable living

Organisations and websites

BioRegional
www.bioregional.com
020 8404 4880
A charity that develops practical sustainability solutions and then sets up new enterprises and partnerships around the world to deliver these solutions on the ground.

Centre for Alternative Technology (CAT)
01654 705950
www.cat.org.uk
CAT offers solutions to environmental problems and demonstrates practical ways of addressing them.

Footprint Trust
01983 822282
www.footprint-trust.co.uk
An Isle of Wight educational charity that works with the Island's community to promote the benefits of sustainable living.

Global Action Plan EcoTeams
020 7836 7345
www.globalactionplan.org.uk/ecoteams-0
The Global Action Plan blends creativity with environmental expertise to help people cut their carbon emissions. Its EcoTeams change community behaviour through education, training and support by providing practical ideas on improving household efficiency, reducing environmental impacts and saving resources.

new economics foundation
020 7820 6300
www.neweconomics.org
An independent think-and-do tank that inspires and demonstrates ways of improving life quality through innovative solutions that challenge mainstream thinking on economic, environmental and social issues.

One Planet Living
www.oneplanetliving.org
A global initiative developed by BioRegional and WWF.

www.oneplanetvision.org/one-planet-living/opl-framework – The One Planet Living framework is made up of 10 principles that help people and organisations to live and work within a fair share of the Earth's resources.

www.oneplanetvision.net – The One Planet Vision website showcases inspiring projects around the world that are using the framework to demonstrate sustainability.

www.oneplanetcommunities.org – One Planet Communities is a network of the Earth's greenest neighbourhoods.

Local groups and regional organisations

This section lists local groups and organisations described in this book.

Bristol Street Trees
www.bristolstreettrees.org
Looks at trends of urban tree removal in Bristol and lobbies to: stop tree removal in public areas, get removed trees replaced and more trees planted, and plan succession planting to balance future tree removal or loss.

CarShareDevon
www.carsharedevon.com
A partnership between Devon County Council and the councils of Plymouth City and Torbay, which aims to maximise people's travel options while also reducing the number of cars on the road.

Edible Islington
020 7527 2000
www.islington.gov.uk/Environment/sustainability/sus_food/edible.asp
Islington Council's campaign for community food growing. It provides grants for new community food-growing projects in the borough.

Greenfinch
01234 827249
www.biogen.co.uk
This project recycles waste food through a low-carbon process in anaerobic digestion plants to make renewable

energy. The remaining liquid is rich in nutrients and used as a biofertiliser.

Hampshire Water Partnership (HWP)
01962 845418
www3.hants.gov.uk/hampshireswater.htm
The HWP was established to enable increased understanding and knowledge-sharing of Hampshire's water environments and the issues facing rivers, wetlands and aquifers.

Incredible Edible Todmorden
www.incredible-edible-todmorden.co.uk
This project increases the amount of local food grown and consumed in the town and involves businesses, schools, farmers and the community.

Low Carbon West Oxford (LCWO)
www.lcwo.org.uk
LCWO was established by residents concerned about climate change and local flooding. It aims to cut carbon emissions by 80% by 2050, encourage residents to live more sustainably and contribute to a more cohesive and resilient community.

Recycling in Cambridge and Peterborough
www.recap.co.uk
This project provides information about recycling schemes, banks and centres, as well as information on swapping or selling your unwanted items.

Recycle Western Riverside (RWR) 'What Not to Waste' makeover
020 7549 0351
www.westernriverside.org.uk
A campaign to encourage residents to recycle, reduce their rubbish and buy more recycled products.

Remade in Brixton
http://remadeinbrixton.wordpress.com
Part of Transition Town Brixton, this is an initiative for zero waste in Britain. It engages residents, schools and businesses to promote local waste reduction, reuse and recycling, and to develop skills in remaking and repair.

Smarter Travel Sutton
www.smartertravelsutton.org
This project aims to improve Sutton by encouraging people to change the types of transport they use, such as cycling or using the buses instead of driving.

Totnes nut tree project
www.totnes.transitionnetwork.org/nuttrees/home
This project aims to provide an extra source of nutritious food for the Totnes community in the future. A film about the project can be seen at www.transitionculture.org/2010/03/01/a-short-film-about-the-totnes-nut-tree-project/

Transition Together
01803 867358
www.transitiontogether.org.uk
A Transition Town Totnes project to enable groups of households to take effective, practical, money-saving and energy-saving steps within their community. Its workbook, *Transition Together*, can be obtained through the Transition Network.

INDEX